Lab Math

A Handbook of Measurements, Calculations, and Other Quantitative Skills for Use at the Bench

Dany Spencer Adams

The Forsyth Institute
Boston, Massachusetts

D1357169

 COLD SPRING HARBOR LABORATORY PRESS
Cold Spring Harbor, New York

LAB MATH

A Handbook of Measurements, Calculations, and Other Quantitative Skills for Use at the Bench

Publisher	John Inglis
Acquisition Editor	David Crotty
Project Manager	Judy Cuddihy
Developmental Editor	Judy Cuddihy
Assistant Developmental Editor	Beth Nickerson
Project Coordinator	Joan Ebert
Production Manager	Denise Weiss
Production Editors	Christine Nolan and Dorothy Brown
Desktop Editor	Susan Schaefer
Interior Designer	Denise Weiss
Cover Designer	Ed Atkeson

Front Cover artwork was created by Jim Duffy (Cold Spring Harbor Laboratory).

Image Credits: p. 79, Modified, with permission, from R.I. Freshney (1994) *Culture of Animal Cells: A Manual of Basic Technique*, 3rd edition (©Wiley-Liss, New York); Cartoons on pages vi, 8, 17, 78, 99, 180, 194, 199, 212, 238, and 243 courtesy of Sidney Harris (copyright held by S. Harris).

Library of Congress Cataloging-in-Publication Data

Adams, Dany Spencer.
  Lab math : a handbook of measurements, calculations, and other
quantitative skills for use at the bench / Dany Spencer Adams.
     p. cm.
Includes index.
  ISBN 0-87969-634-6 (alk. paper)
  1. Mathematics--Laboratory manuals. I. Title.
  QA40.A34 2003
  510--dc21
                                     2003012916

10 9 8 7 6 5 4

All Cold Spring Harbor Laboratory Press publications may be ordered directly from Cold Spring Harbor Laboratory Press, 500 Sunnyside Boulevard, Woodbury, New York 11797-2924. Phone: 1-800-843-4388 in Continental U.S. and Canada. All other locations: (516) 422-4100. FAX: (516) 422-4097. E-mail: cshpress@cshl.edu. For a complete catalog of all Cold Spring Harbor Laboratory Press publications, visit our World Wide Web Site http://www.cshlpress.com/

For Bert

"OH, DEAR— ANOTHER TRAGIC CASE OF MATH ANXIETY!"

Contents

WHAT'S YOUR FAVORITE TIP/TECHNIQUE?

While every attempt has been made to fill this book with as much good, practical information as possible, I know that there must be other great shortcuts, techniques, equations, and explanations that have not been included. Because some of the omissions were made due to ignorance of their existence, I invite you to submit your comments, your favorite math method, or the topic you wish had been covered.

Please include in your proposal:

1. The trick/technique/equation/explanation you wish to share or would like to see explained or tabulated.

2. An attribution (if possible).

3. A fully worked example (if appropriate).

4. A brief explanation of why you think it should be included in *Lab Math*.

5. Information for contacting you.

Please help us to make the second edition of *Lab Math* even better by sending your comments and suggestions to:

LabMath@cshl.edu

Thank you very much in advance for your input,

Dany Spencer Adams

Preface

THIS BOOK IS ABOUT THE NUMBERS YOU MAY READ, rely on, discover, and report during your time in the lab. It is written for three audiences. The first audience is comfortable with mathematics, but will find it useful to have reminders and references collected in one place. That audience will probably keep this book on a shelf above the bench. The second audience is right in the middle of something and needs to know immediately how to accomplish a numerical task, such as calculating the grams per liter to achieve a particular molarity. That audience has this book open on the bench top. The third audience wants, and has time for, an explanation of where the numbers and equations come from. That audience probably has this book on their desk. (There may be a fourth audience that just reads the cartoons.) All users are enthusiastically encouraged to be part of any or all of these audiences, as circumstances warrant.

I began writing and collecting the information in this book because I had gotten tired of explaining to students how to convert units, calculate molarity, create a recipe for a solution, use a pH meter, calibrate an ocular micrometer, etc. I'm sure you have your own list. I had also begun posting notes to myself about the things I just couldn't seem to remember, such as how many grams per milliliter make a 10% solution. I expect there are a few notes posted around your lab, too. Word of my growing collection of How Tos and Short Cuts for working with numbers began to spread, and soon colleagues were telling me about the mathematical topics that they just couldn't remember or that they were tired of teaching, and I started adding new sections by request. Eventually, the information outgrew the Cell Biology lab manual where I was "publishing" it, and I began to have it assembled on its own as the Laboratory Calculations Reference Manual. It soon became *Lab Math*, although, all along, I've thought it could be titled "What I Wish I Could Remember from Algebra, Geometry, and Introductory Biology, Physics, and Chemistry." Once I began to get requests from professors wanting to use the collection in their own labs and classrooms, I decided it was time to go public. When I discovered Kathy Barker's *At the Bench*, I decided to approach Cold Spring Harbor Laboratory Press; happily, they said yes.

Chapter 1 includes some basics about numbers themselves and some generic types of measurements. This is the chapter that is most relevant to everything else; becoming comfortable with the information in Chapter 1 will make all numbers easier, and time spent learning it now will save untold hours later. Chapter 2 covers the numbers used to describe the chemistry that underlies much biology. Like Chapter 1, this chapter contains fundamentals that, if learned now, will save you time and hassle later. Chapter 3 describes how to calibrate and correctly use equipment commonly found in biological laboratories. I recommend two uses for this chapter. First, make sure that your equipment is being used properly, and, where appropriate, perform and post calibrations. Second, have newcomers to the lab read the sections appropriate for your equipment as a prelude to hands-on training. Chapter 4 gives methods and short cuts for making solutions. In my experience as a teacher, I have found this to be the most useful chapter in the book. Chapters 5 and 6 cover methods used in Molecular Biology. Chapter 7 comprises an introduction to statistics and includes advice on how to communicate numerical data. Chapter 8 contains reference tables and charts that you probably already have somewhere in your bookshelf, but now you know exactly where to look first. It also contains "Plug and Chugs," blank equation forms for some of the more common numerical tasks. These are ready to be copied in your laboratory notebook.

The people at Cold Spring Harbor Laboratory Press, especially Dave Crotty, Beth Nickerson, Judy Cuddihy, Joan Ebert, Dotty Brown, Susan Schaefer, and Mala Mazzullo have been terrific; my thanks go to them for their skill, their intelligence, and their unfailing kindness. Many people reviewed different sections and provided useful critiques and ideas; my sincerest thanks to each of them. I owe a great deal to the Helen Riaboff Whiteley Center at the Friday Harbor Laboratories. A three-week fellowship there allowed me to focus and write most of the first version of the manuscript; I cannot imagine a better or more beautiful place to work. I am grateful to the students at Smith College who were guinea pigs for early versions of certain sections; they were happy to let me know what was good as well as absolutely delighted to let me know about mistakes. The support and encouragement of colleagues, friends, and family have been wonderful. Particular mention goes to Melanie Adams, Jeanne Powell, Rachel Fink, and Steven Beeber, each of whom helped in incalculable ways. Finally, I wish to acknowledge the fundamental contributions of teachers, especially Jonesy Wagner, Margot Gumport, Michael Sturm, Yves Volel, M.A.R. Koehl, Tom Daniel, and Garry Odell; thanks to all of you for everything, but especially for knowing all along that girls like math.

Dany Spencer Adams

Acknowledgments

THE FOLLOWING PEOPLE PROVIDED invaluable support, including suggesting topics, reading drafts, actually *using* early versions in their labs and classes, pre-ordering copies, providing soy lattes and places to work, and cheering me on when I needed it.

James R. Adams

Deborah L. Chapman

Stacey Combes

The Forsyth Institute

David Grosof

Jennifer Innes

Sandra Laney

Michael Levin

Michelle Lizotte-Waniewski

Helen McBride

Mary Murphy

Patricia Parsons-Wingerter

Jennifer Pinkham

Lori Saunders

Abby Silvan

Ulrike Spaete

Starbucks

Christine White-Ziegler

CHAPTER 1

Numbers and Measurements in the Laboratory

Scientists count and measure. So, if you do science, it is practically inevitable that you will work with numbers, and it helps to be comfortable both reading and manipulating numerical data. Unfortunately, many people are not comfortable with numbers. In fact, the very idea of working with numbers can generate anxiety, not to mention pages of random calculations that often reflect a vain hope of conjuring the correct answer by chance. To make things worse, if you work with numbers, you work with uncertainty. There is no way around it: No measuring device is perfect; *all* measurements contain some error. That means you also have to be comfortable with measures of uncertainty, i.e., numbers that describe your numbers. Luckily, there are straightforward methods for working with numerical data, for minimizing error, and for being candid about uncertainty; they all are essential to good science, and they all make sense. To help you develop your own sense of the methods used to gather, analyze, and report numerical data, this first chapter explains why scientists report numbers the way they do.

TALKING ABOUT NUMERICAL DATA

Dimension: A property or physical characteristic of something. There are seven fundamental (or base) dimensions: amount of substance (N), length (L), mass (M), time (T), temperature (Θ); electric current (A), and luminous intensity (J). Other dimensions, such as concentration, volume, and voltage, are derived from combinations of these seven. As indicated in parentheses, the symbols for dimensions are written in uppercase Roman letters (with the exception of temperature which is an uppercase Greek Theta). These symbols are, however, rarely used. (See also Chapter 8, Table of SI Units.)

Equation: In mathematics and in this book, an equation is a statement indicating that the expression on the left-hand side equals the expression on the right-hand side. In chemistry, an equation is an expression describing a chemical reaction.

Magnitude: The size or value of a measurement. Measurements usually have both magnitude and dimensions. The incubator might be at 37°C; the magnitude is 37 and the dimension is temperature (°C is the unit; see below).

Mantissa: The word "mantissa" has more than one meaning. In this book, mantissa refers to the number left when 10 to some exponent has been factored out to write a number in scientific notation. For example, in the number 6.02×10^{23}, 6.02 is the mantissa. Computer programmers also use the word mantissa.

Order of Magnitude: This term indicates the relative size of a number and is often used in comparisons of two numbers. For example, "same order of magnitude" means that two numbers are about the same; "different order of magnitude" means that they are very different. Specifically, it refers to the power of ten that can be factored out of the number: 7.8×10^5 is two orders of magnitude larger than 7.8×10^3.

Significant Digits: The digits in a number that contain meaningful information about the value, but not the order of magnitude, of the number. Scientific notation is the best way to report the correct number of significant digits. The number of digits in the mantissa is the number of significant digits. (See also below, How Many Significant Digits?)

Unit: An agreed-upon standard for a dimension. Système Internationale (SI) units are the current standard and are used throughout this book. The SI units of the seven fundamental dimensions are: for amount of substance, moles (mol); for length, meters (m); for mass, kilograms (kg); for time, seconds (s); for temperature, Kelvin (K); for electric current, Amperes (A); and for luminous intensity, candelas (cd). As indicated in parentheses, the symbols for units are lowercase Roman letters, with the exception of K and A, which are uppercase because these units were named for people. (See also Chapter 8, Table of SI Units.)

Unit Fraction: A fraction in which the numerator equals the denominator (it therefore has a magnitude of 1), but in which different units appear. For example, 1 inch/2.54 cm is a unit fraction, as is 1000 mm/100 cm. Unit fractions are useful for converting among units because they describe the relationships among the different units that appear; they are conversion factors.

REPORTING MEASUREMENTS

How a scientist reports a measurement tells more about that measurement than just a numerical value. Consider the following:

4.50×10^{-2} m

The numerical value being reported is 0.0450; but there is a lot more information in that expression. It contains information about the uncertainty of the measurement as well as information about what was measured.

What Was Measured: Units

The units of a measurement tell you what was measured; sometimes the units contain information about the magnitude of the value. In this example, the unit is meters (m). This tells you that a length was measured, because the accepted unit for the dimension length is meters. This value could also have been reported as 45.0 millimeters (mm), where the prefix "milli" on the unit "meter" provides information about the order of magnitude of the value.

Uncertainty

The uncertainty of a measurement is indicated within the format of the number itself. In this example, the uncertainty is not indicated explicitly, but you know it is there (it is *always* there). By convention, nothing explicit means ±1 in the last digit. In this case, it means that the scientist is certain the length is between 44.9 mm and 45.1 mm (some people interpret nothing explicit as meaning ±.5 in the last digit, i.e., 45.0 means 44.5 to 45.5). If the number is reported as $4.55 \times 10^{-2} \pm 0.05 \times 10^{-2}$ m, the uncertainty is reported explicitly: It is the number after the ±, i.e., 0.05×10^{-2} m. In this case, the scientist believes the value of the length to lie between 4.50×10^{-2} and 4.60×10^{-2} m.

Two phenomena contribute to uncertainty: inaccuracy and imprecision. To minimize uncertainty, you want to maximize both accuracy and precision.

- *Accuracy* is a measure of how close a measurement is to the true value. An accurately thrown dart is close to the bull's-eye. To maximize accuracy, make sure that the measuring device is properly calibrated; to quantify and report accuracy, calculate and report a statistic, such as the standard deviation, that describes the spread of the data (see Chapter 7).

- *Precision* is a measure of repeatability. Precisely thrown darts are clumped together (although they may be nowhere near the bull's-eye). To maximize precision, use the best instrument you can afford; to quantify and report precision, determine the limits of the measuring device and report the measure using the correct number of significant digits (see below).

These two terms are often used interchangeably, and sometimes they are confused. In particular, the word "accuracy" is often used to describe what is, by the above definition, precision. The confusion may arise because it is unlikely (although not impossible) that a particular machine would be accurate but imprecise; therefore, if one is claiming accuracy, precision is implied. The converse is not true. Many machines are precise but inaccurate, which is why they must be calibrated. The definitions given above reflect the most commonly stated definitions and distinctions; however, good sources give competing explanations. When in doubt, just say "error" or "uncertainty," both of which imply vagaries of both accuracy and precision. In this book, the above definitions will be used.

Both accuracy and precision are affected by the measuring device and experimental technique.

☑ Examples

- *Measuring device/scientist combination A:* One scientist warms up and calibrates a machine and then measures the concentration of a 4.355 μM solution ten times and reports $c = 4$ μM, which indicates that this measuring device gave ten concentration values between 3 μM and 5 μM for the solution. This is an accurate measure, but it is not very precise.

- *Measuring device/scientist combination B:* A second scientist takes the same 4.355 μM solution and measures its concentration ten times using a machine that has not been calibrated. The scientist reports $c = 5.5$ μM. This measure is more precise (the ten measurements varied only by ±0.1, compared with ±1 for device A), but it is not accurate.

- *Measuring device/scientist combination C:* A third scientist warms up a third machine for 15 minutes, calibrates it, measures the solution ten times, and reports a value of 4.35 μM. This measurement is accurate (the true value, 4.355 μM, is within the range indicated by the measure, i.e., 4.34 to 4.36 μM), and it is the most precise of the three (the measurements varied by only ±0.01). More importantly, it is precise *enough* for the scientist's purposes.

All of the scientists in this example are measuring the same solution. The actual concentration isn't changing. What is changing is the accuracy and precision of the measuring device being used. Whether a measurement is accurate enough and precise enough is up to the researcher.

The precision of the measurement is indicated within the format of the number itself. Specifically, is it indicated by the number of significant digits, and the number of significant digits is best indicated by using scientific notation (although scientific notation is only *required* if the number reported ends in one or more zeroes).

SCIENTIFIC NOTATION

Scientific notation is a way to write numbers that takes advantage of the powers of ten. For example, the number 2,843,651 is equal to 2.843651 x 1,000,000, which is equal to 2.843651×10^6, which is scientific notation. We factor out a power of ten because our number system is a base-10 number system. (One reason the metric system is preferred by scientists is that it is also based on tens.)

Mantissas and Exponents

Names are given to the different parts of a number written in scientific notation. In the example 2.843651×10^6, the 2.843651 is called the mantissa and the 6 is the exponent. The 10 is said to be raised to a power of 6. In scientific notation, the number raised to a power is always 10. So, to use scientific notation, it is useful to be familiar with powers of ten.

Powers of Ten

The equations below show a variety of ways of writing the powers of ten between 10^3 (1000) and 10^{-3} (0.001) If you scan these numbers, you should be able to get a sense of how easy it is to go from one form to the other. You can also see why $10^0 = 1$ and why a negative exponent means 1 over 10 to the positive exponent.

$$10^3 = 1000 = \frac{1000}{1} = \frac{10^3}{1}$$

$$10^2 = 100 = \frac{100}{1} = \frac{10^2}{1}$$

$$10^1 = 10 = \frac{10}{1} = \frac{10^1}{1}$$

$$10^0 = 1 = \frac{1}{1} = \frac{10^0}{1}$$

$$10^{-1} = 0.1 = \frac{1}{10} = \frac{1}{10^1}$$

$$10^{-2} = 0.01 = \frac{1}{100} = \frac{1}{10^2}$$

$$10^{-3} = 0.001 = \frac{1}{1000} = \frac{1}{10^3}$$

Notice that each time you divide by 10, the power goes down by 1, and each time you multiply by 10, the power goes up by 1. This makes it very simple to take a number in scientific notation and multiply or divide it by 10. When you multiply a number in scientific notation by 10, you are adding 1 to the exponent on the 10; this is the same as moving the decimal in the mantissa one place to the right. Dividing by 10 means subtracting 1 from the exponent, or moving the decimal in the mantissa one place to the left.

To convert from a number written conventionally to a number written in scientific notation, move the decimal point to the left or the right until there are one or two nonzero digits to the left of the decimal point. Then count the number of digits you passed along the way, and that is the exponent on the 10. If you move to the left, the exponent is positive; if you move to the right, the exponent is negative. For example, move the decimal point to the right:

$$0.000256734 \rightarrow 00.00256734 \rightarrow 000.0256734 \rightarrow 0000.256734 \rightarrow 00002.56734$$

The mantissa is 2.56734 and, since you moved the decimal four times to the right, the exponent is -4, making the number 2.56734×10^{-4} in scientific notation.

☑ Examples: Scientific Notation

Example	Original Number	Same Number in Scientific Notation
1	64,600	6.46×10^4
2	0.0004009	4.009×10^{-4}
3	25,000,000	2.500×10^7
4	25,000,000	2.5×10^7

The examples above show how to convert numbers into scientific notation. In Example 1, the power of ten that can be factored out to leave behind a real number with one digit to the left of the decimal is 10,000 or 10^4. Using the shortcut, you move the decimal four places to the left to transform 64,400 into proper scientific notation. In Example 2, factoring out 10^{-4} leaves behind a real number with one nonzero digit to the left of the decimal. Using the shortcut, you move the decimal place four places to the right. In Examples 3 and 4, 10^7 can be factored out to result in proper scientific notation.

Examples 3 and 4 also demonstrate how scientific notation allows you to indicate the precision of the measuring apparatus even if the values end in zeroes. For example, if 25,000,000 was reported as the number of bacteria on an agar plate, it should imply that some graduate student (the measuring device) counted every bacterium individually, many times, and the number of bacteria counted varied from 25,000,001 to 24,999,999. You would have to wonder if this was reported correctly. If the number of bacterial cells was reported as in Example 3, 2.500×10^7, you could infer that the counts were all $\pm 0.001 \times 10^7$, i.e., 1×10^4—still iffy, but more believable. If the number of cells was reported as in Example 4, 2.5×10^7, you would know that the value reported is $\pm 1 \times 10^6$, which is definitely believable. And if the count were truly precise to eight significant digits (I can't even come up with a possible scenario), it should be reported as 2.5000000×10^7.

Manipulating Expressions in Scientific Notation

The key to easily multiplying expressions in scientific notation is to remember that multiplication is commutative. In other words, order does not matter (the numbers can commute between positions). Thus:

$$(a * b) * (c * d) = (a * c) * (b * d)$$

To make multiplication that looks complicated simple, rearrange the numbers:

$$4.1 \times 10^2 * 3.2 \times 10^4 = (4.1 * 3.2) * (10^2 * 10^4) = 13 \times 10^6$$

In simple terms, multiply the mantissas and add the exponents. To divide numbers in scientific notation, divide the mantissas and subtract the exponents:

$$(4.1 \times 10^2) \div (3.2 \times 10^4) = (4.1 \div 3.2) \times (10^2 \div 10^4) = 1.3 \times 10^{-2}$$

Remember that dividing is the same as multiplying by 1 over the number; therefore:

$(4.1 \times 10^2) \div (3.2 \times 10^4)$ is the same as $(4.1 \times 10^2) \times 1/(3.2 \times 10^4)$ or $\dfrac{4.1}{3.2} \times \dfrac{10^2}{10^4}$

which equals 1.3×10^{-2}.

Recall that, when no power of ten is written, the power of ten is 10^0 ($10^0 = 1$). Thus:

$$\frac{5.678 \times 10^3}{4.12} = 1.38 \times 10^3$$

because

$$\frac{5.678 \times 10^3}{4.12 \times 10^0} = \frac{5.678}{4.12} \times \frac{10^3}{10^0} = 1.38 \times 10^3$$

"I'M AFRAID THAT'S NOT THE WAY TO BRIDGE THE COMMUNICATIONS GAP BETWEEN SCIENTISTS AND NON-SCIENTISTS."

Reasons to Use Scientific Notation

There are three reasons to use scientific notation.

1. Scientific notation often allows you to write, in just a few digits, numbers that would otherwise have unwieldy lengths. Think of numbers that have lots of zeroes, like Avogadro's number: 602,214,199,000,000,000,000,000. This number is almost always written 6.02×10^{23}. (*Note:* it could be written more precisely as $6.02214199 \times 10^{23}$.)

2. It is very easy to do computations with numbers written in scientific notation, especially if all you need is an estimate. For example, suppose you need to know about how much you have of a solution. In the freezer, you find 23 aliquots of 1500 µl each. You could calculate the amount longhand or try to find a calculator, but you don't really

need an exact figure—you just need to know whether to reorder the solution. It is much simpler to estimate: 23 is about 10^1 and 1500 is about 10^3, so I have about 10^4 μl.

3. As explained above (Examples 1, 3, and 4), scientific notation is the best way to represent the precision of a measurement if the value of the measurement begins or ends with zeroes. (If a number begins and ends with any other digit, you can just count the number of digits; they are all significant.)

PRACTICALITIES OF SIGNIFICANT DIGITS

How Many Significant Digits?

A digit is significant if it conveys information about the value of the number, but not if it *only* contains information about the *order of magnitude* of the number. Digits 1 through 9, therefore, are always significant, because they can only tell you about value. Zero, however, is confusing, because a zero might be significant; but it might just be indicating another multiple of ten. The rules about 1–9 and about 0s are described below.

Which Digits Are Significant?

Rule I: All nonzero digits are significant.
Rule II: Zeros between nonzero digits are significant.
Rule III: Zeros to the right of the first nonzero digit are significant.
Rule IV: Zeros to the left of the first nonzero digit are NOT significant.

Number	In Scientific Notation	Number of Significant Digits	Relevant Rules
4998.3040	4.9983040×10^3	8	I, II, III
0.000351	3.51×10^{-4}	3	I, IV

Reading measurements

To determine the precision of someone else's measurement based on the number of significant digits reported, you must assume that the researcher has reported the number correctly. If the number ends with a nonzero digit, just count the nonzero digits, and that is the number of significant digits. If the number ends with zero, you can only be sure about the number of digits if the number is reported in scientific notation. If it is, you can count the significant digits: It is the number of digits in the mantissa. If the number is not reported in scientific notation, the number of significant digits may be less obvious.

For example, assume you are reading a report about a survey of birds on a small island, and the report says there are 13,000 little brown birds in a square mile of the island. It is highly unlikely that the scientists counted every single bird in a square mile and then repeated the measurements. It is also highly unlikely that, even if they did, the mean came out to be exactly 13,000. What they probably did was count the number of birds in some randomly chosen smaller areas and then multiply the mean to get an estimate of the total number in a square mile. If the scientists reported the number using two significant digits—1.3×10^4 birds mi^{-2} (birds per square mile)—you would know that the actual census numbers, after taking means and multiplying, all fell between 12,000 and 14,000. If, however, the scientists' calculated counts of little brown birds all fell between 13,200 and 13,400, they could report the number as 1.33×10^4, three significant digits. The number of significant digits in the second example reflects a more precise count than that cited in the first report; these counts varied by only ± 100, whereas the first counts varied by ± 1000. By being careful about the correct number of significant digits, you convey important information (the precision or repeatability) of your measurement. And, the best way to be clear about the number of significant digits is to report your numbers in scientific notation.

Reporting measurements

To determine how many significant digits to report for a particular measurement, you must determine the precision of the measuring device. To determine the precision, either repeatedly measure the same thing and note the range of values reported by the measuring device or look it up in the documentation that came with the instrument. Then report the measurement in scientific notation to convey the precision to the reader.

☑ Example

You need to keep track of the temperature of the seawater in which sea urchin embryos are developing. In this example, two different measuring devices are used. Measuring device 1 is a mercury thermometer, with black lines indicating increments of 0.1°C.

Can you tell if repeated measures vary between 10°C and 11°C? **Yes.**
Can you tell if repeated measures vary between 10.5°C and 10.6°C? **Yes.**
Can you tell if repeated measures vary between 10.57°C and 10.56°C? **No.**

Thus, it would be overstating the precision (lying about your uncertainty) to report a temperature of 10.56°C—the best you can say is 10.6°C. You should therefore report your temperature as 10.6°C or as 1.06 x 10^{1}°C; each clearly has three significant digits.

Measuring device 2 is a digital thermocouple, with a readout that shows three decimal places. You could look up the precision of the thermocouple in a manual, or you could do the same test as in the exercise above:

Can you tell if repeated measures vary between 10.56°C and 10.57°C? **Yes.**
Can you tell if repeated measures vary between 10.565°C and 10.566°C? **Yes.**
Can you tell if repeated measures vary between 10.5656°C and 10.5657°C? **No.**

In this case, 10.566°C is the best you can say, or 1.0566 x 10^{1}°C (five significant digits). The thermocouple has higher precision than the mercury thermometer.

What Happens to Significant Digits When You Do Arithmetic?

When scientific measurements are multiplied or divided, the result has the same number of significant digits as the measurement with the fewest significant digits. For example:

$$1.440 \times 10^{-6} \div 5.66609 = 2.540 \times 10^{-7}$$
$$4 \qquad\qquad\quad 6 \qquad\qquad\quad 4$$

When scientific measurements are added or subtracted, the result should have the same number of significant digits to the right of the decimal place as the measurement with the fewest decimal places. Examples are:

$$200 + 4.56 = 205$$
$$0 \quad\;\; 2 \qquad 0$$

$$1.440 \times 10^{-2} - 5.6 \times 10^{-5} = 1.434 \times 10^{-2}$$
$$5 \qquad\qquad\;\; 6 \qquad\qquad\quad 5$$

The second example above illustrates that it may be easier to count decimal places if the number is not written in scientific notation.

Ignore exact numbers when determining the precision of the calculated value. Some values have no uncertainty; e.g., 1 liter is exactly 1000 milliliters. These numbers do not affect the precision of a calculated value; so ignore them when determining significant digits. In the following example, NA indicates not applicable:

$$1.440 \times 10^{-2} \frac{mg}{ml} \times 10^{3} \frac{ml}{l} = 1.440 \times 10^{1} \frac{mg}{l}$$
$$4 \qquad\qquad\qquad\quad NA \qquad\quad 4$$

When reporting a value, use the above rules to determine how to report the precision of your measurement, and use scientific notation to unambiguously indicate the number of significant digits.

Prefixes and Significant Digits

Sometimes you can use a prefix on a unit of measurement, instead of scientific notation, when reporting a measurement. Just be sure to report the correct number of significant digits.

☑ Example

Phospholipid bilayers are reported to be about 7 nm thick. That same length could be written as 7×10^{-9} m. In this example, the 10^{-9} is represented by the accepted prefix nano- or "n" which means "$\times 10^{-9}$." The length does not change, the magnitude of the measurement does not change, and the number of significant digits does not change.

☑ Example

Someone reported using 1.00×10^{-5} liters of solution in an experiment. The volume could have been reported as 10.0 μl ($1.00 \times 10^{-5} = 10.0 \times 10^{-6}$, and the prefix that means "$\times 10^{-6}$" is micro- or μ). In this example, however, it is not clear whether those two zeroes are significant or not. The volume could also be rewritten as 1.00×10^{1} μl, but it is common practice to neglect scientific notation if the exponent on the 10 would be –1, 0, 1, or 2. So, in fact, 1.00×10^{-5} is the best choice.

Summary: Precision, Scientific Notation, and Significant Digits

- You must report the precision of your scientific measurements so that readers of your publications can properly interpret your experiments. If you overstate your precision, you risk reporting misleading numbers and, even worse, reaching incorrect conclusions.

- The precision of your measurements is determined by the measuring devices you use.

- The best way to report precision is to report the correct number of significant digits. If your measurement begins or ends with zeroes, report it in scientific notation.

- To indicate the magnitude of a measurement, you can maintain the proper number of significant digits but replace scientific notation with an appropriate prefix on the unit of measurement.

DIMENSIONS AND UNITS

Dimensions

Dimensions are the physical properties you measure in the laboratory. The seven fundamental dimensions are amount of a substance (N), mass (M), length (L), time (T), temperature (Θ), electric current (A), and luminous intensity (J). Fundamental dimensions can be combined to give derived dimensions, such as volume (L^3), and force (MLT^{-2}).

Some properties are dimensionless, e.g., radians, ratios, percents, and Reynolds number.

A given property *always* has the same dimensions, but it can be described using different units.

Units

A unit is an agreed-upon standard for a dimension. Much of the scientific world has agreed that the Système Internationale (SI) units is the accepted standard. The official brochure documenting SI units (The International System of Units [SI], 7th ed. 1998. Bureau International des Poids et Mesures) can be downloaded at www.bipm.fr. Another resource for information about SI units is physics.nist.gov.

Units provide information about the variables that have been measured, and if used with appropriate prefixes, they may also provide information about the magnitude of the measurement. The concentration of a solution, for example, might be reported as 1 M (1 molar or 1 mole per liter), and this tells you that there are a certain number of molecules per liter of fluid. If the reported concentration of a dilution of this solution units is 1 mM (1 millimolar or 1 millimole per liter), this indicates that the concentration is a thousand times smaller than the concentrated solution.

The agreed-upon units for the seven fundamental dimensions are moles (mol) for amount of a substance, kilograms (kg) for mass, meters (m) for length, seconds (s) for time, kelvins (K) for temperature, amperes (A) for electric current, and candelas (cd) for luminous intensity. Units of fundamental dimensions are combined to give the units for the derived dimensions. For example, volume is length cubed, so the resulting units for volume are meters cubed [m^3]; force is mass multiplied by length divided by time squared, so the unit of force is kg m s^{-2} or newtons (N). An extensive table of dimensions, units, and symbols is presented in Chapter 8.

Other units and unit systems are still used and are found in older literature. A common one is CGS (centimeter-gram-second). For a very useful collection of factors for converting among different units, see C.J. Pennycuick's *Conversion Factors: SI Units and Many Others*, *The CRC Handbook of Chemistry and Physics*, or use one of the online converters.

Converting between different units

Despite the existence of an accepted standard for units of measurement, different units are often used to describe the same dimension: 1.00 inch is equivalent to 2.54 centimeters; 10 kilometers is equivalent to 10^4 meters. Often, it is necessary to convert the units that you have into the units that you want. To do so, you simply need to know the relationships between the units. These relationships are referred to as conversion factors. You can look conversion factors up in tables (see below), but it is worth memorizing the ones that are most relevant to your work.

☑ Examples: Conversion Factors

1 foot = 0.3048 meters
1 kilogram = 2.2046 pounds
1 gallon = 4.54609 liters

Below is a method for converting any unit into any other unit (provided the dimensions stay the same; you cannot convert meters into amperes, no matter how hard you try). This method is based on the following two truths:

1. Units in an equation, like numbers, can be cancelled out.

2. Because a unit fraction is a fraction whose numerical value is 1, you can multiply by as many unit fractions as you want.

METHOD Converting Units

The basic idea is to multiply an expression by unit fractions as needed to convert the starting units into the ones you need. This will not change the value of the measurement, because you are just multiplying by 1. It will merely change the units of the measurement, which is what you want to do. Here are the steps:

1. Write the conversion as an equation:

    starting number (starting units) = new number (new units)

2. Count the number of different units in the equation.

3. Add that many empty fractions to the left-hand side of the equation.

4. Start to fill in those fractions by adding whatever units you need to cancel the old units and adding whatever units you need to make the new units appear.

5. Make the fractions equal to 1 (turn them into unit fractions).

6. Multiply.

☑ **Example**

Escherichia coli are about 1 μm (1 micron) in length. How long are they in picometers?

1. Write down the conversion as an equation:

   $1 \ \mu m \ = \ X \ pm$

2. Count the number of different units that appear in the equation: micrometers and picometers; one, two.

3. On the left-hand side, draw as many empty fractions as you have units.

   $1 \ \mu m \ \times \ \left(\dfrac{\quad}{\quad}\right) \times \left(\dfrac{\quad}{\quad}\right) \ = \ X \ pm$

4. Start to fill in the fractions so that the cancellations you need can happen and so that correct new units will appear.

   $1 \ \mu m \ \times \left(\dfrac{\quad}{\mu m}\right) \times \left(\dfrac{pm}{\quad}\right) \ = \ X \ pm$

5. Turn those fractions into unit fractions by looking for relationships you know. You know, or you can find out, from consulting the Prefixes for Units table in Chapter 8 that 1 μm = 10^{-6} m and 1 pm = 10^{-12} m. This information can be used as follows:

   If 10^{-6} m $=$ 1 μm, then $\dfrac{10^{-6} \ m}{1 \ \mu m}$ is a unit fraction.

   and

   If 10^{-12} m $=$ 1 pm, then $\dfrac{10^{-12} \ m}{1 \ pm}$ is a unit fraction.

6. Now, fill in the equation:

$$1\ \mu m\ \times\ \frac{10^{-6}\ m}{1\ \mu m}\ \times\ \frac{1\ pm}{10^{-12}\ m}\ =\ X\ pm$$

or, equivalently:

$$1\ \mu m\ *\ \frac{1\ m}{10^{6}\ \mu m}\ *\ \frac{10^{12}\ pm}{1\ m}\ =\ X\ pm$$

Now, cancel both μm's and m's and solve:

$$1\ \mu m\ =\ 10^{6}\ pm$$

This method is a basic guideline for converting units. This example used two of many possible unit fractions to convert micromoles to picomoles. When doing unit conversions, it does not matter which, or how many, unit fractions you use, as long as your conversion factors are correct.

EQUATIONS

An equation is a statement indicating that two expressions are equal. This information can be extremely useful if you are an experimental scientist.

Reasons Why Equations Are Extremely Useful

There are three reasons why equations are extremely useful:

1. If an equation tells you one quantity equals another, then you do not have to measure both. You can choose what parameters in an equation you want to measure and derive the ones you want to calculate.

   Consider an equation that everybody knows: $E = mc^2$. This equation tells you that if you want to know the energy of something, all you have to do is measure its mass (m), and then do a simple multiplication. Why is this useful? Consider that to measure energy (E) directly, you would have to find a bomb calorimeter, put a known mass of the substance into the sample bomb, supply heat (Q) to the sample and to a reference (the difference is ΔQ), and subtract the amount of heat required to heat the bomb when empty (Q_0, also known as the heat equivalent). The heat capacity is $c_v = (\Delta Q - \Delta Q_0)/m\Delta T$. Then comes the hard part: calculating the uncertainty, which comes from the measurement of ΔT, the change of volume work adjustment, the calibration of the platinum thermometer, radiation from the thermometer head, and drift in the ice point resistance. You also have to consider the reproducibility of the measurements. There is more, but you get the point.

2. Equations define the relationships among variables. The more complicated the research subject, the more you can appreciate how thorny it can become trying to understand and follow the relationships among all the different phenomena that may affect your experimental system. An equation sums it all up in one clean expression. You can write an equation on a laboratory whiteboard, and it will be a constant source of reassurance and direction in your research.

3. Because equations tell you two things are equal, they also tell you that the dimensions on the two sides of the equals sign must also necessarily be equivalent. This simple truth gives rise to the easiest way to do a preliminary check of your work, that is, check that the units on both sides of an equation balance. It also leads to the useful, some might say astoundingly useful, technique of dimensional analysis.

Dimensional Analysis

The most commonly used method that comes from dimensional analysis is balancing units. Making sure the units balance is a simple way to do a first check of calculations. The units on the left-hand side must equal the units on the right-hand side. If they don't, there is something wrong and you must go back and figure out what it is. Examples of how to check that units balance are given in Chapter 4.

"LADIES AND GENTLEMEN, OUR RESEARCH DEPARTMENT HAS COME UP WITH THIS. WHAT DO WE WISH TO DO WITH IT?"

If two expressions are equal, which is exactly what an equation tells you, then their units must be equal as well. (Joules do not equal candelas, so any equation that says they do *must* be wrong.) You can use that rule to figure out what might, and what could not, affect your phenomenon. That is, you can use the absolute requirement for balanced units (called homogeneity of dimensions) to predict which variables determine the value of your measurement. This allows you to write an equation that is both a testable hypothesis and a guide to what needs testing.

The following are some examples where we already know the equation that describes the phenomenon.

☑ Example A

The classic example is the equation describing the period of a simple pendulum (a bob at the end of a string). The answer is:

$$t = 2\pi\sqrt{\frac{l}{g}}$$

where:

t = period (seconds per cycle)	dimension	= T
l = length of the pendulum	dimension	= L
g = acceleration due to gravity	dimensions	= LT^{-2}

If you didn't know that (or couldn't look it up), how could dimensional analysis help you?

1. Think about what variable quantities might affect the period of a simple pendulum. Make a list of the candidates:

   • The mass of the bob (m; dimension M)

   • The weight of the bob (f; dimensions MLT^{-2})

   • The length of the string (l; dimension L)

   • The radius of the bob (r; dimension L)

   • Gravity (g; dimensions LT^{-2})

   This is not a comprehensive list of all the quantities you might propose. This is one of the limitations of the technique: There is no criterion for choosing the most suitable quantities. Still, the worst that usually happens is that you come up with a trivial hypothesis, rather than an erroneous one. In any case, if you start with this list and don't do dimensional analysis, you will have to do four experiments (five if you can get space on the shuttle) to determine which of these variables actually affects the period.

Now look at the list and see if you can reduce it by making some simplifying assumptions. Simplifying assumptions, when used with great care, are very helpful. For this example, it seems reasonable to assume that, compared to the length of the string, the radius of the bob is negligible. (For now, we will proceed on that assumption; in reality, you should do the experiments to test your assumptions.)

2. Write the first version of the equation/hypothesis: The period (t; dimension T) is a function of the mass, the force, the length, and gravity:

   $t = f(m, f, l, g)$

3. The goal of the whole process is to determine what f looks like; here is where the dimensions come in. We start from the fact that any function can be expressed as a series of the following form:

   $t = C_1 \times m^a f^b l^c g^d + C_2 \times m^{a'} f^{b'} l^{c'} g^{d'} + C_3 \times m^{a''} f^{b''} l^{c''} g^{d''} + \text{etc.}$

   where the Cs are unknown coefficients. Now, we know that each variable must always have the same dimensions, thus, $a = a' = a''$ and $b = b' = b''$, etc. Thus, the equation simplifies to:

   $t = C \times m^a f^b l^c g^d$

   where:

   $C = C_1 + C_2 + C_3 + \text{etc.}$

   We also know that the dimension of t is T; so the dimensions of $C \times m^a f^b l^c g^d$ must also be equal to T (homogeneity of dimensions).

4. Rewrite the equation, substituting the dimensions of the variables for the variables themselves:

   $T = (C)(M)^a (MLT^{-2})^b (L)^c (LT^{-2})^d$

5. The term $(M)^a$ can drop out because the way this is written, a change to the mass of the bob (M^a) *alone* would have to be reflected in a change to the period (T). But there is no M term on the left-hand side, i.e., the M term is not balanced; there cannot be a lone M term on the right-hand side. a must equal zero. The dimension equation now looks like this:

   $T = (C)(MLT^{-2})^b (L)^c (LT^{-2})^d$

6. Consolidate the dimensions so that each dimension appears only once:

$$T = (C)(M^b)(L^{b+c+d})(T^{-2b-2d})$$

7. Now there are three unknowns, b, c, and d; dimensional homogeneity gives the three equations needed to solve for their values. The exponent b must equal 0; there are no M terms on the left-hand side. There are also no L terms on the left-hand side; so $b + c + d = 0$ too. Also, $-2b - 2d$ must equal 1, because the exponent on the T on the left-hand side is 1.

$b = 0$ (there are no M terms on the left-hand side)
$b + c + d = 0$ (there are no L terms on the left-hand side)
$-2b - 2d = 1$ (there is T^1 on the left-hand side)

Solve however you like: Simple substitutions are fine; matrices are good.

$-2(0) - 2d = 1$
$-2d = 1 - 0$

$2d = -1$
$d = -1/2$

$0 + c - 1/2 = 0$
$c = 1/2$

Now, go back to the dimension equation, which now looks like this:

$$T = (C)(MLT^{-2})^0 (L)^{\frac{1}{2}} (LT^{-2})^{\frac{-1}{2}}$$

8. Substituting back the variables gives:

$$t = (C)(1)^{\frac{1}{2}}(g)^{\frac{-1}{2}}$$

which can be rewritten:

$$t = C\sqrt{\frac{1}{g}}$$

which is very close to the actual equation. All that is missing is the value of C, which must be determined experimentally. What this did was narrow down the number of necessary experiments to one. If you vary l, you can determine the value of C (which turns out to be 2π). If, at this point, you ended up with a trivial equation, such as $t = t$, go back and redo the analysis without dropping terms at step 5, or by dropping different terms.

This equation is a hypothesis about the relationship of *l* to *t*, and it directs your experiments because it tells you that varying *l* should cause *t* to vary in a very particular way. If you do the experiment, you will either discover the value of C or discover that the left-hand side does not equal the right-hand side, thus disproving the hypothesis. If that happens, go back and reexamine your simplifying assumptions.

☑ Example B: Dropping Stones from the Leaning Tower of Pisa

Galileo believed that the mass of a stone was irrelevant to the distance traveled in a given amount of time. He demonstrated this by dropping stones of different masses from the leaning tower of Pisa and showing that they landed simultaneously. His opposition felt otherwise, choosing, inexplicably, to disbelieve their eyes. If the anti-Galileos had used dimensional analysis, they would have proved that Galileo was correct:

1. Think about and list what might matter. Simplify if possible. The distance traveled (*d*; dimension L) should depend on:

   - The mass of the object (*m*; dimension M)
   - The amount of time spent falling (*t*; dimension T)
   - Gravity (*g*; dimensions LT^{-2})
   - No simplifying assumptions

2. Write the starting function:

   $d = f(m,t,g)$

3. Convert to series:

   $d = C \times m^a t^b g^c$

4. Substitute dimensions:

   $L = C \times (M)^a(T)^b(LT^{-2})^c$

5. Drop unbalanceable variables: M^a can drop out; there is no way to reflect a change in M on the left-hand side. Right here, the anti-Galileos would have seen the error of their ways.

6. Consolidate fundamental dimensions:

   $L = (C)(L^c)(T^{b-2c})$

7. Solve for exponents:

$$c = 1$$
$$b - 2c = 0$$
$$b = 2$$

8. Substitute variables and exponents:

$$d = (C)(t)^2(g)^1 = Ct^2g$$

The experiment would show that the value of C is $1/2$; $d = 1/2\ gt^2$. Galileo was correct.

Notes

The point of this is not to avoid doing all experiments; it is to avoid doing unnecessary or redundant experiments.

The above two examples were rigged. In both cases, there were enough equations to solve for all the unknowns. It may not come out that way in practice. To work through a case where there are more variables than fundamental dimensions, you have to work with the math a bit more and know something about your system.

Summary of Dimensional Analysis

1. Think about and list quantities that might matter. Simplify if possible.
2. Write the starting function.
3. Convert to $f(J,K,L,M) = C(J)^s(K)^t(L)^u(M)^v$.
4. Substitute dimensions.
5. Drop unbalanceable terms.
6. Consolidate fundamental dimensions.
7. Solve for exponents.
8. Replace variables. Don't forget C.

Be wary: There is no criterion for selecting the most suitable quantities. You cannot predict the values of coefficients. You may lose track of dimensionless variables.

MEASURING FUNDAMENTAL PROPERTIES

There are seven properties of matter that are defined as "fundamental" properties (also sometimes referred to as primary or basic properties). These are: amount of a substance, length, mass, time, temperature, current, and luminous intensity. They are considered fundamental because they are completely independent of one another. That is, there is no way to derive one from any others. Each of these fundamental properties has its own symbol and its own SI unit:

Fundamental Property	Symbol for Property	Dimension	SI Unit	Symbol for Unit
Amount of substance	n	N	mole	mol
Length	l, d	L	meter	m
Mass	m	M	kilogram	kg
Time	t, τ	T	second	s
Temperature	T	Θ	Kelvin	K
Electric current	I	A	Ampere	A
Luminous intensity	I_v	J	candela	cd

All other properties, the "derived properties," are combinations of fundamental properties. Refer to the table in Chapter 8 for more details. See also the SI units brochure listed in the Resources at the end of this chapter.

Amount of a Substance

Dimension: N.

SI Unit: Moles (mol).

Physical Basis: A mole is the number of atoms in 12 grams of carbon-12 (^{12}C). That number is 6.02×10^{23}; so there are 6.02×10^{23} individuals in a mole of any entity. If you are talking about moles of elements or molecules, the molecular weight is the mass in grams of one mole of that element or molecule.

Equipment: Balance and calculator.

To determine the number of moles of a substance empirically, you would have to be able to count atoms, something most laboratories cannot do. Instead, you use the relationship of moles to molecular weight, explained above, to derive the number of moles of a substance from the substance's mass. Simply put, to determine the amount, or moles, of a chemical, you determine its mass on a balance, then divide by its molecular weight (i.e., multiply by moles per gram). That gives you the number of moles of the chemical.

When the unit mole is used, the elementary entities must be specified; they may be atoms, molecules, ions, electrons, other particles, or specified groups of such particles. Typically in the laboratory, you would say "I added 1 mole of sodium chloride," thereby defining the entity.

When reporting smaller values, you just report the property (n) without a unit, rather than as a fraction of a mole. For example, you would report $n = 602$ blue colonies, not $n = 1 \times 10^{-21}$ (i.e., $602 \div 6.02 \times 10^{23}$) moles of blue colonies.

If you are counting a large number of items, such as the number of cells in a culture, it can help to have a counter. Counters are little devices that keep track of the number of times you have pushed a button. All you have to do is push the button once for every item you count.

Length

Dimension: L.

SI Unit: Meters (m).

Physical basis: A meter is defined as the distance light travels in a vacuum in 1/299,792,458th of a second. The length of a meter was originally defined relative to the size of the Earth (one ten-millionth part of a quadrant of the Earth) in 1799. It was later redefined as the length of a bar of platinum-iridium that was stored in a vault in Sèvres, France. The newest definition allows anyone on Earth, who happens to have spectacularly sophisticated equipment, to measure a meter with fantastic accuracy.

Equipment: Ruler and caliper.

To measure the length of something, you compare it to something whose length you know, usually a ruler. A ruler is a stick or strip that has been scored at regularly spaced, standardized intervals. Rulers come in different sizes and different units. You choose a ruler based on the approximate size of what you are measuring and on the precision with which you need to measure. To measure lengths in the range of inches, you use a 6- or 12-inch ruler. To measure lengths in yards, you use a yardstick. (To measure lengths in miles, you would most likely use an odometer.) To measure lengths in the range 10 mm to 1 m, you use a 150-mm ruler, a 300-mm ruler, or a meter stick. To measure lengths in the range of 0 to 150 mm, you use calipers.

To convert other units of length into meters, use the following conversion factors:

UNIT If you have this unit	CONVERSION FACTOR do this to the number	to get the SI unit
Cubit	x 4.318 x 10^{-1}	= meters
Fathom	x 1.8288	= meters
Foot	x 3.048 x 10^{-1}	= meters
Furlong	x 2.01168 x 10^2	= meters
Hand	x 1.016 x 10^{-1}	= meters
Inch	x 2.54 x 10^{-2}	= meters
League	x 4.828042 x 10^3	= meters
Mil	x 3 x 10^{-5}	= meters
Mile	x 1.609344 x 10^3	= meters
Nautical league	x 5.556 x 10^3	= meters
Nautical mile	x 1.852 x 10^3	= meters
Yard	x 9.144 x 10^{-1}	= meters

Mass

Dimension: M.

SI Unit: Kilogram (kg).

Physical Basis: A kilogram is based on the international prototype of kilogram.

Equipment: Balance.

Newton defined mass as the "quantity of matter." There is no definition of mass yet that is based on fundamental or atomic constants. Mass is often confused with weight and density, both of which are discussed below in Measuring Derived Properties. Briefly, density is mass of substance per unit of volume. Weight is mass times gravitational acceleration, making weight a force. One way to remember the difference between mass and weight is to remember that the mass of an object stays constant, even when that object is on the moon; however, the same object would weigh less on the moon than on Earth because the moon has less gravity.

To determine the mass of an object or substance, measure its weight and divide by acceleration due to gravity. Most balances do the division for you, so the number reported on the readout is the mass, as indicated by the units printed on the screen. You can also use a double-beam balance to determine mass. A double-beam balance, which is a sophisticated version of the device held aloft by the blindfolded goddess Justice, lets you determine mass by comparing your objects to objects of known mass. To convert among units of mass, use the following conversion factors:

UNIF If you have this unit	CONVERSION FACTOR do this to the number	to get the SI unit
Carat	$\times$ 2 $\times$ 10^{-4}	= kilograms
Grain	$\times$ 6 $\times$ 10^{-5}	= kilograms
Metric ton (tonne)	$\times$ 10^3	= kilograms
Ounces (Avoirdupois)	$\times$ 2.835 $\times$ 10^{-2}	= kilograms
Ounces (Troy)	$\times$ 3.11 $\times$ 10^{-2}	= kilograms
Pennyweight	$\times$ 1.56 $\times$ 10^{-3}	= kilograms
Poundal	$\times$ 1.409 $\times$ 10^{-2}	= kilograms
Pounds (Avoirdupois)	$\times$ 4.536 $\times$ 10^{-1}	= kilograms
Pounds (Troy)	$\times$ 3.7325 $\times$ 10^{-1}	= kilograms
Scruple	$\times$ 1.3 $\times$ 10^{-3}	= kilograms
Slug	$\times$ 1.460592 $\times$ 10^1	= kilograms
Stone	$\times$ 6.3504	= kilograms

Time

Dimension: T.

SI Unit: Second (s).

Physical basis: A second is defined as the amount of time it takes for a [133]Cesium atom to transition between two hyperfine levels of the ground state 9,192,631,770 times.

Equipment: Clock or timer.

To determine how long until lunch, you use a clock that is divided into hours and minutes. To determine how long it takes for an Olympic runner to run a mile, you use a clock that is divided into minutes, seconds, and fractions of seconds. To determine how long it has been since your fossilized plant was alive, you use the decay of ^{14}C as a clock.

As with any measuring device, you choose a clock based on the approximate magnitude of the result and on the desired precision and accuracy of the result. For most applications in a biology laboratory, the wall clock or a timer that keeps track of time to the nearest second is sufficient.

To convert other units of time to seconds, use the following conversion factors:

UNIT If you have this unit	CONVERSION FACTOR do this to the number	to get the SI unit
Minute (sidereal)	$\times$ 5.983617 $\times$ 10^1	= seconds
Minute	$\times$ 6.0 $\times$ 10^1	= seconds
Hour	$\times$ 3.600 $\times$ 10^3	= seconds
Day	$\times$ 8.6400 $\times$ 10^4	= seconds
Week	$\times$ 6.04800 $\times$ 10^5	= seconds
Fortnight	$\times$ 1.2096 $\times$ 10^6	= seconds
Month	$\times$ 2.628 $\times$ 10^6	= seconds
Year	$\times$ 3.1536 $\times$ 10^7	= seconds
Decade	$\times$ 3.15576 $\times$ 10^8	= seconds
Century	$\times$ 3.1556736 $\times$ 10^9	= seconds
Millennium	$\times$ 3.15569088 $\times$ 10^{10}	= seconds

Temperature

Dimension: Θ.

SI Unit: Kelvin (K).

Physical basis: The kelvin is the fraction 1/273.16 of the thermodynamic temperature of the triple point of water. 0 K (= –273.16°C) is absolute zero; absolute zero is the temperature below which it is impossible to go.

Equipment: Thermometer and radiometer.

Thermometers vary in their sophistication. The difference between thermometers is reflected in the accuracy and precision of the readings they provide, the units they measure, the range of temperatures they can measure, and the cost. Such information is typically provided in scientific catalogs as well as in the paperwork that comes with a thermometer. The following is a summary of the different categories of temperature-measuring devices.

Bulb or liquid-in-glass thermometer

This thermometer is a long, thin cylinder, with a fluid-filled bulb connected to a thin, sealed tube. The fluid in a bulb thermometer is most often mercury or a red alcohol that is easy to see. Bulb thermometers work because fluids expand when heated; the geometry of the interior space (relatively large volume in the bulb versus a relatively small volume in the tube) makes the change in volume due to heating of the inner liquid visible. The degree of expansion of the inner liquid is converted into a measure of temperature. In some thermometers, there is, between the bulb and the tube, a narrowing called a bore. While heating moves mercury through the bore and into the tube, cooling is not sufficient to move the liquid back through the bore into the bulb. That is why you must shake a mercury thermometer to get the mercury back into the bulb.

Bimetallic strip thermometer

In this type of thermometer, two different types of metals (usually iron and brass) are welded together. When the temperature rises, the two metals expand by different amounts, causing the strip to bend. The extent of the bend is converted into a measure of temperature.

Electrical thermometers

Resistance temperature detectors (RTDs) are designed on the principle that the resistance of a metal changes with a change in temperature. The measure of resistance in the metal is converted to a measure of temperature. Platinum is often used in RTDs because its temperature–resistance relationship is linear over a large range. RTDs are found in appliances that heat or cool and in electronic circuits, motors, and other electronic devices.

A thermistor is an electrical resistance thermometer made with a ceramic material instead of metal wires. The resistance of the ceramic varies with temperature and, as with RTDs, the resistance is converted into a measure of temperature. Digital thermometers used by health care practitioners use thermistors that are very accurate, but only over a small range.

A thermocouple is based on the principle that the voltage generated between two dissimilar metals joined in at least one position is a function of temperature. By using different combinations of metals, thermocouples can be designed to measure a wide range of temperatures. Thermocouples are used in many devices.

Radiometer

A radiometer measures the temperature-dependent radiation (usually infrared) at a particular wavelength as it is emitted from a particular substance. The amount of radiation is converted to a measure of temperature. To convert among different units of temperature, use the following conversion factors:

Converting to Kelvin

K = °C + 273

$$K = (°F - 32) \times \frac{5}{9} + 273$$

Converting to Celsius

°C = K − 273

$$°C = (°F - 32) \times \frac{5}{9}$$

Converting to Fahrenheit

$$°F = (K - 273) \times \frac{9}{5} + 32$$

$$°F = (°C \times \frac{9}{5}) + 32$$

Electric Current

Dimension: A.

SI Unit: Ampere (A).

Physical Basis: The ampere is that constant current which, if maintained in two straight parallel conductors of infinite length or negligible circular cross-section, and placed 1 meter apart in vacuum, would produce between these conductors a force equal to 2×10^{-7} newton per meter of length. (An ampere is the amount of current that would be produced by an electromotive force of 1 volt acting through a resistance of 1 ohm.)

Equipment: Ammeter.

Current and voltage are often confused. Current is the *rate* of flow of electric force in a conductor from a point of higher potential to a point of lower potential. Voltage is the electromotive *force* created by the difference in electrical potential.

Classic ammeters employ galvanometers to measure current by siphoning off a known fraction of the current in the circuit being tested. Within the galvanometer, deflection of a needle surrounded by an electrical field will be proportional to the current passing through the meter. That deflection is converted to a measure of current. Newer ammeters can directly measure the voltage across a tiny length of wire that is carrying current. That measure of voltage is converted into a measure of current using Ohm's law: $I = V/R$ (current = voltage/resistance).

Luminous Intensity

Dimension: J.

SI Unit: Candela (cd).

Physical Basis: The candela is the luminous intensity, in a given direction, of a source that emits monochromatic radiation of frequency 540×10^{12} hertz and that has a radiant intensity in that direction of 1/683 watt per steradian.

Equipment: Radiometer.

Luminous intensity is measured in candelas. Radiometers measure lux, which is lumens per area. Lumens per steradian gives candelas. (A steradian is to a sphere as a radian is to a circle. If you think of a radian as defining a triangle, you can think of a steradian as defining a cone.) To convert from lux to luminous intensity, you need to position the equipment so that you know the number of steradians and the area over which the light is being measured. How to do this is described in the instructions for your instrument.

The business end of a radiometer is the light sensor. Light sensors are designed to cover a particular band of wavelengths (e.g., visible light) and a range of magnitudes. The magnitude is quantified as the irradiance, which has units of watts per meter squared, but which are reported on the radiometer as $mW\ cm^{-2}$.

MEASURING DERIVED PROPERTIES

A derived property is one whose dimensions are combinations of fundamental dimensions. The derived properties mentioned below are commonly measured in biology laboratories. There are many, many more derived properties. For more details, see Chapter 8.

Area

Dimensions: L^2.

SI Unit: Square meter or meters squared (m^2).

Equipment: Ruler, digitizing pad, and computer.

How to measure the area of a shape or object depends a great deal on what you are measuring, as described below.

Geometric shapes in two or three dimensions

Use a ruler to measure the key length or lengths required, then use equations from geometry to calculate the area of the whole shape. For example, to measure the area of a rectangle, multiply the length of the short side by the length of the long side to obtain the area of the surface. The area of a cube is calculated by measuring the area of one side and multiplying by 6 (six sides of a cube). For equations used to calculate the area of many common shapes, see Chapter 8.

Flat, irregular shapes

To measure the area of an irregular shape with no associated equation, cut out a paper silhouette of the shape, weigh it, then convert to area (weight x area per weight of the paper). This is cheap and easy, but it has a lot of error associated with it.

Another method to measure irregular shapes is to use a computer program to print a grid onto a transparency. Typically, you place the grid over a shape, count nodes, and then use statistics to determine the area. This is a good way to compute area because you will have a measure of the uncertainty of your value. There are computer programs that will tell you the area of a shape when it is traced in. NIH Image (see Resources) is one that is both useful and free.

Three-dimensional, irregular shapes

If an oddly shaped item such as a fossil or a fish is evenly coated with a dye, the amount of dye that sticks to the surface will be proportional to the surface area. You can determine the amount of dye by weighing the object before and after dipping or by rinsing the dye into a known volume of fluid and using a spectrophotometer. The amount of dye can be converted to surface area by means of a previously determined area:weight or area:absorbance equation. You can determine these relationships by measuring the dye stuck to objects with known surface areas.

To convert between different units of area, use the following conversion factors:

UNIT If you have this unit	CONVERSION FACTOR do this to the number	to get the SI unit
Acre	x 4.046873 x 10^3	= meters^2
Ares	x 10^2	= meters^2
Foot^2	x 9.29 x 10^{-2}	= meters^2
Hectare	x 10^4	= meters^2
Inch^2	x 6.5 x 10^{-4}	= meters^2
League^2	x 2.3309990 x 10^7	= meters^2
Mile^2	x 2.58998811 x 10^6	= meters^2
Rod^2	x 2.529296 x 10^1	= meters^2
Rood	x 1.01171825 x 10^3	= meters^2
Township	x 9.3239940 x 10^7	= meters^2
Yard^2	x 8.3613 x 10^{-1}	= meters^2

Volume

Dimensions: L^3.

SI Unit: Cubic meters or meters cubed (m^3).

Equipment: Graduated cylinder and balance.

The volume of something is the amount of space it occupies in three dimensions. To measure the volume of a fluid, pour it into a volumetric container, such as a graduated cylinder. You can get a good estimate of the volume of a small amount of fluid by picking it up with a pipetter and then dialing the pipetter until the fluid reaches, but does not exit, the tip of the pipette. To measure the volume of a solid you can do either of the following:

1. Measure how much fluid it displaces when submerged in a volumetric container.

2. Weigh it and divide by the density of the material.

To convert between different units of volume, use the following conversion factors:

UNIF If you have this unit	CONVERSION FACTOR do this to the number				to get the SI unit
Ale gallon (UK)	× 4.62	= liter	× 10^{-3}	= meters^3
Barrel (crude oil)	× 1.589873 × 10^2	= liter	× 10^{-3}	= meters^3
Barrel (liquid, US)	× 1.1924048 × 10^2	= liter	× 10^{-3}	= meters^3
Board foot	× 2.35974	= liter	× 10^{-3}	= meters^3
Bushel	× 3.523907 × 10^1	= liter	× 10^{-3}	= meters^3
Chaldron	× 1.26861 × 10^3	= liter	× 10^{-3}	= meters^3
Fluid ounce	× 2.957 × 10^{-2}	= liter	× 10^{-3}	= meters^3
Foot^3	× 2.831685 × 10^1	= liter	× 10^{-3}	= meters^3
Gallon (dry, US)	× 4.40488	= liter	× 10^{-3}	= meters^3
Gallon (liquid, US)	× 3.78541	= liter	× 10^{-3}	= meters^3
Gallon (UK)	× 4.54609	= liter	× 10^{-3}	= meters^3
Hogshead	× 2.384806 × 10^2	= liter	× 10^{-3}	= meters^3
Inch^3	× 1.639 × 10^2	= liter	× 10^{-3}	= meters^3
Peck	× 8.80977	= liter	× 10^{-3}	= meters^3
Pint (dry, US)	× 5.5061 × 10^{-1}	= liter	× 10^{-3}	= meters^3
Pint (liquid, US)	× 4.7318 × 10^{-1}	= liter	× 10^{-3}	= meters^3
Quart (dry, US)	× 1.10122	= liter	× 10^{-3}	= meters^3
Quart (liquid, US)	× 9.4365 × 10^{-1}	= liter	× 10^{-3}	= meters^3
Tun	× 9.5392 × 10^2	= liter	× 10^{-3}	= meters^3
Yard^3	× 7.645549 × 10^2	= liter	× 10^{-3}	= meters^3

Rate, Velocity

Dimensions: LT^{-1}.

SI Unit: Meters per second (m s^{-1}).

Equipment: Ruler, clock, and graph paper; speedometer.

Rate and velocity are both a change in some property in a certain amount of time; thus, the SI units always include "per second" [s^{-1}]. Velocity is used to describe a change in position over a change in time, that is, meters per second. Rate is used more generically, as in reaction rate (change in concentration over time) or mortality rate (number of deaths over time). To determine the rate or the velocity, you graph the values of the property that is changing (e.g., position or concentration) on the *y* axis, and time on the *x* axis. The slope of the resulting curve is the rate or the velocity. When you take the slope, you are measuring the change in the property over the change in time (remember "rise over run?"). (*Note:* The technical difference between a rate and a velocity is that rate is a scalar, whereas velocity is a vector; that is, rate has magnitude, whereas velocity has magnitude and direction.)

Acceleration

Dimensions: LT⁻².

SI Unit: Meters per second per second, or meters per second squared (m s⁻²).

Equipment: Accelerometer.

Acceleration is the measure of a change in velocity. Acceleration caused by gravity on Earth, for example, is 9.8 m/s², that is, an object at sea level is being accelerated downward at a velocity of 9.8 m/s². Accelerometers are based on the principle that acceleration will cause a deflection in a material. The deflection of the material (the sensor) is converted to an electrical signal (this can be accomplished in a variety of ways), and that electrical signal is converted to a measure of acceleration. As with most measuring devices, accelerometers are available in a variety of sensitivities, precisions, sizes, and costs. Some accelerometers can only detect acceleration in a single direction; others can keep track simultaneously in three directions. Accelerometers report acceleration in "gs." To convert from gs (which, technically, is not a unit) to m s⁻², multiply by 9.8.

Density

Dimensions: ML⁻³.

SI Unit: Kilogram per cubic meter (kg m⁻³).

Equipment: Graduated cylinder and balance.

Density is the ratio of the mass of an object to its volume. If you have two cubes of equal volume, one made of sponge and the other of lead, the lead cube will have a greater density because the mass of lead is greater than that of sponge.

To determine the density of a solid, you measure the volume and determine the mass; then you divide the mass by the volume:

$$\rho = \frac{m}{V}$$

where:

$$\rho = \text{density} \left(\frac{kg}{m^3} \right)$$

m = mass [kg]
V = volume [m³]

To convert between different units of density, use the following conversion factors:

UNIT If you have this unit	CONVERSION FACTOR do this to the number	to get the SI unit
lbm foot⁻³	× 1.602 × 10¹	= kilograms per meter³
lbm gallon⁻¹	× 1.19824 × 10²	= kilograms per meter³
Slug foot⁻³	× 5.153788 × 10²	= kilograms per meter³
Long ton yard⁻³	× 1.328939 × 10³	= kilograms per meter³
Short ton foot⁻³	× 3.2040 × 10⁴	= kilograms per meter³

lbm = pound mass.

Weight

Dimensions: MLT^{-2}.

Units: Newtons (N).

Equipment: Balance.

The weight of an object is its mass times gravitational acceleration. This means that weight is a force, so its units are Newtons. To determine the weight of something, you measure its mass using a balance and then multiply the mass (in kg) times gravity (9.81 ms⁻²). If your balance reads 1.403 g, the weight is 1.38×10^{-2} N (because 1.403 g = 1.403×10^{-3} kg and 1.403×10^{-3} kg × 9.81 ms⁻² = 1.38×10^{-2} N).

Pressure

Dimensions: $M\ L^{-1}T^{-2}$.

SI Units: Pascals (Pa).

Equipment: Pressure gauge or pressure transducer.

Pressure is force exerted against an opposing body. Pressure is measured in pascals, but it is frequently expressed as N m⁻², or weight per area, which is more intuitive.

Pressure gauges come in a variety of types. The basic components can measure different ranges of pressures with different degrees of precision. Because pressure comes from many different places, there are also different categories of pressure transducers, some of which have their own names:

- Barometer measures atmospheric pressure.

- Manometer measures vacuum pressure.

- Sphygmomanometer measures blood pressure.

All of these instruments convert the displacement of something (a fluid, usually) into a measure of pressure. There are many different units of pressure; very few devices give readings in pascals.

To convert other units of pressure to pascals, use the following conversion factors:

UNIT If you have this unit	CONVERSION FACTOR do this to the number	to get the SI unit
Atm	$\times$ 1.01325 $\times$ 10^5	= pascals
Bar	$\times$ 10^5	= pascals
cm H_2O (4°C)	$\times$ 9.80638 $\times$ 10^1	= pascals
cm Hg (0°C)	$\times$ 1.33322 $\times$ 10^3	= pascals
dyne cm^{-2}	$\times$ 10^{-1}	= pascals
ft H_2O (39.2°F)	$\times$ 2.98898 $\times$ 10^3	= pascals
ft seawater	$\times$ 3068167 $\times$ 10^3	= pascals
Gram force cm^{-2}	$\times$ 9.80665 $\times$ 10^1	= pascals
in H_2O (39.2°F)	$\times$ 2.49082 $\times$ 10^2	= pascals
in H_2O (60°F)	$\times$ 2.4884 $\times$ 10^2	= pascals
in Hg (32°F)	$\times$ 3.386389 $\times$ 10^3	= pascals
in Hg (60°F)	$\times$ 3.37685 $\times$ 10^3	= pascals
kgf cm^{-2}	$\times$ 9.80665 $\times$ 10^4	= pascals
kgf m^{-2}	$\times$ 9.80665	= pascals
kgf mm^{-2}	$\times$ 9.806650 $\times$ 10^6	= pascals
KSI (Kip in^{-2})	$\times$ 6.894757 $\times$ 10^6	= pascals
lbf ft^{-2}	$\times$ 4.788026 $\times$ 10^1	= pascals
Meters seawater	$\times$ 10066345 $\times$ 10^4	= pascals
Poundal ft^{-2}	$\times$ 1.48816	= pascals
psi (lbf in^{-2})	$\times$ 6.894757 $\times$ 10^3	= pascals
Torr (0°C)	$\times$ 1.33322 $\times$ 10^2	= pascals

kgf = kilogram force; lbf = pound force; Kip = 1000 x lbf.

Voltage

Dimensions: M L^2 T^{-3} A^{-1}.

SI Units: Volts (V).

Equipment: Voltmeter.

Voltage and resistance are intimately related to the fundamental property current. This relationship is described by Ohm's law, $I = V/R$.

Voltage (V) is electromotive force or an electrical potential. Most meters that measure electrical characteristics can measure volts, amperes, ohms, and farads. Classic voltmeters employ galvanometers to measure current by siphoning off a known fraction of the current in the circuit being tested. Within the galvanometer, deflection of a needle surrounded by an electrical field will be proportional to the current passing through the meter. That deflection is converted to a measure of current, which is then converted into a measure of voltage using Ohm's law. Newer voltmeters can directly measure the voltage across a tiny length of wire that is carrying current.

RESOURCES

Conversion Factors

Pennycuick C.J. 1988. *Conversion factors: SI units and many others.* The University of Chicago Press, Chicago, Illinois.

On-line Calculators for Unit Conversions

http://www.megaconverter.com

SI System

The International System of Units (SI). Bureau International des Poids et Mesures, Sèvres, France. (http://www.bipm.fr)

Taylor B.N. 1995. *Guide for the use of the international system of units (SI).* NIST special publication 811. National Institute of Standards and Technology, Washington, D.C. (http://www.physics.nist.gov/Document/sp811.pdf)

http://www.physics.nist.gov

CHAPTER 2

Chemistry by the Numbers

MATTER

The entities that make up matter, namely, molecules and atoms, are largely understood in terms of quantifiable characteristics such as mass and charge. So, understanding chemistry often means understanding numbers. Like all disciplines, chemistry has its own vocabulary, which can be confusing and it is often misused.

Talking about Atoms and Molecules

Atom: The smallest unit or particle of an element that has all the properties of that element. It consists of a nucleus containing positively charged particles called protons and neutral particles called neutrons; orbiting the nucleus are negatively charged particles called electrons.

Atomic Mass (Atomic Weight): The mass (also referred to as weight, although it is not technically a weight) of one atom of an element is equal to the combined weight of the protons, neutrons, and electrons of that atom (approximately the same as the combined mass of the protons and neutrons, as electron weight is almost negligible). The atomic mass of an element listed on the periodic table (see p. 45) is either a weighted average of the masses of all the isotopes of that atom or the mass of the most common isotope. It is reported in "atomic mass units" (amu) (abbreviation m_u) or daltons (abbreviation D, or sometimes, incorrectly, Da). One mole of any element has a mass in grams equal to one atom's mass in amu. For example, the atomic mass of sodium is 22.99. One atom of sodium has an atomic mass of 22.99 m_u or 22.99 D; one mole of sodium has a mass of 22.99 g. To prove this, do a quick calculation:

22.99 D/atom $\times$ 1.6605 $\times$ 10^{-24} g/D $\times$ 6.0221 $\times$ 10^{23} atoms/mole = 22.99 g/mole

Atomic Mass Unit (amu): One-twelfth the mass of a ^{12}C nucleus; used to describe atomic and formula weights.

1 amu = 1 dalton = $1.6605387 \times 10^{-27}$ kg

Atomic Number: The number of protons in the nucleus of one atom of an element. Atomic number defines an element. For example, if the nucleus of an atom contains 19 protons, by definition it is an atom of potassium.

Atomic Weight (Atomic Mass): The mass of one atom of an element. It is reported in "atomic mass units" (amu) (abbreviation m_u) or daltons (abbreviation D, or sometimes, incorrectly, Da).

Avogadro's Number: A constant, A, equal to $6.02214199 \times 10^{23}$. It is the number of particles, for example, molecules, in one mole of a substance. For example, there are $6.02214199 \times 10^{23}$ molecules of EDTA in a mole of EDTA. Similarly, there are $6.02214199 \times 10^{23}$ molecules of water in a mole of water.

Avogadro Constant: A constant, N_A, equal to $6.02214199 \times 10^{23}$ mole^{-1}. The constant is easily confused with Avogadro's number; the difference is the constant has units of mole^{-1}.

Dalton: One-twelfth the mass of a ^{12}C nucleus. For example, one molecule of bovine serum albumin has a mass of 66×10^3 D, or 66 kD.

1 Dalton = 1 atomic mass unit

Decay: What happens when an atom loses particles to become another element. ^{238}U decays to ^{234}Th (emitting an α particle). ^{234}Th decays to ^{234}Pa (emitting a β particle).

Electron: Subatomic particle with a mass of $0.000548\ m_u$ and a charge of -1. Electrons are located in orbits around the nucleus of an atom. Electron orbits are described by a probability function.

Element: Atom or atoms of a particular atomic number. If an atom has an atomic number of 14, it is an atom of the element silicon.

Free Radical: An uncharged chemical species that has one or two unshared (or unpaired) electrons, i.e., electrons in the outermost shell (or valence electrons) that are not participating in a chemical bond. Free radicals are highly reactive and enthusiastically form chemical bonds that result in the filling of the outermost shell. OH is a free radical called the hydroxyl free radical.

Ion: An atom that has lost or gained one or more electrons, resulting in a change in the charge of the atom. Cl$^-$ (chloride) is an ion of chlorine; it has one more electron than Cl. Ca^{2+} is an ion of calcium; it has two fewer electrons than Ca.

Anion: A negatively charged ion. F^- is an anion.

Cation: A positively charged ion. Na^+ is a cation.

Isotopes: Different forms of a single element. Different isotopes of an element differ in atomic weight due to differences in the number of neutrons in the nucleus. Lead has four naturally occurring isotopes: ^{204}Pb, ^{206}Pb, ^{207}Pb, and ^{208}Pb. Pb always has 82 protons; ^{204}Pb has 122 neutrons, ^{206}Pb has 124 neutrons, ^{207}Pb has 125 neutrons, and ^{208}Pb has 126 neutrons.

Molar: An adjective that describes the concentration (in moles per liter) of a solution. For example, a solution containing 2.5 moles of NaCl per liter of water is said to be "a 2.5 molar salt solution."

Molar Mass: The mass of one mole of a substance. The molar mass of water is 18.015 g (the sum of the mass of two moles of hydrogen plus the mass of one mole of oxygen).

Mole: A name for 6.022×10^{23} particles of something (just like "dozen" is a name for 12 particles of something). One mole of a chemical has a mass equal to the molecular weight in grams. For example, β-estradiol has a molecular weight of 272.39; so one mole of β-estradiol "weighs" (has a mass of) 272.39 grams. Mole is also the SI unit for amount of substance.

Molarity: Moles of solute per liter of solvent (for more information, see Chapter 4, "Making Solutions"). The molarity of a solution is the number of moles per liter in that solution; the 2.5 molar salt solution mentioned above has a molarity of 2.5.

Molecule: A collection of characteristic atoms bound together in a characteristic way. Water is a molecule with two atoms of hydrogen bound to one atom of oxygen. The structure of any molecule is determined by the atoms' charges and sizes. Physically, the water molecule resembles a V, with the O in the middle and the two Hs on either side.

Neutron: Subatomic particle with a mass of 1.00866 m_u and a charge of 0. Neutrons are located in the nucleus of an atom.

Nuclides: Generic name for any form of any element. A list of all nuclides would include all the isotopes of all the elements. The symbol for a nuclide is written $^A_Z E$. where A is the mass number (the number of protons plus the number of neutrons), Z is the atomic number (the number of protons), and E is the symbol for the element. For example, $^{204}_{82}Pb$ is a nuclide, as are $^{206}_{82}Pb$ and ^2_1H.

Proton: Subatomic particle with a mass of 1.007276 m_u and a charge of +1. It is located in the nucleus of an atom. The number of protons in an atom is the atomic number of the atom, and it determines the identity of the element. If the nucleus of an atom has 17 protons, it is an atom of chlorine.

Radioactivity: Emission of energy (i.e., radiation) specifically due to the decay of an atomic nucleus. The energy is emitted in one or more of the following forms: alpha (α) particles, beta (β) particles, and gamma (γ) rays. Radioactivity is ionizing radiation; i.e., the radiation emitted during the decay of a radioactive element is energetic enough to displace electrons (thus creating ions) from another substance upon interaction.

Subatomic Particles: The proton, the neutron, and the electron.

Particle	Symbol	Charge	Mass	Mass
Proton	p^+	+1	1.67265×10^{-27} kg	1.007276 D
Electron	e^-	−1	9.10953×10^{-31} kg	0.000548 D
Neutron	n	0	1.67495×10^{-27} kg	1.00866 D

Valence: The combining ability of an atom, which is determined by the number of electrons in the outermost shell, also called the valence shell. These electrons are called the valence electrons. Hydrogen has a valence of 1, because when it forms a bond, it shares one electron with another atom. Oxygen has a valence of 2. Carbon has a valence of 4.

Molecular, Empirical, and Formula Weight

Empirical Formula: The lowest whole-number ratio of atoms in a compound; e.g., the empirical formula for hydrogen peroxide (molecular formula H_2O_2) is HO. This term is also used when describing ionic compounds, such as NaCl, where the smallest units of the substance are ions, not molecules.

Formula Weight (FW): Also called the empirical molar mass, this refers to the mass of the empirical formula, which, as indicated above, tells you only the relative proportion of each atom and thus is useful for talking about ionic compounds. The formula weight of sodium chloride is 58.44 (because Na^+ has a mass of 22.99 and Cl^- has a mass of 35.45). The units of formula weight are g $mole^{-1}$ and, because they are always the same, they are not usually written out.

Most people say MW regardless of what the compound is. Chemical companies put "FW" on their labels; what FW refers to in this case is the weight of the formula as listed on the label. So, for example, if it is a hydrated compound, the FW includes the mass of the water molecules.

Molecular Formula: This term applies to molecules and describes the number of atoms of each element within a single molecule. $C_6H_{12}O_6$ is the molecular formula of glucose; one molecule of glucose has 6 carbon atoms, 12 hydrogen atoms, and 6 oxygen atoms.

Molecular Weight (MW): Also called the molecular molar mass, this refers to the mass of the molecular formula, which tells you both the number of atoms of each element and their relative proportions in the molecule. The molecular weight of glucose is 180.2. The units of molecular weight are g mole^{-1} and, because they are always the same, they are not usually written out. This term is now sharing space with "relative molecular weight" (M_r).

*Relative Molecular Weight (**M**$_r$):* This is the molecular weight of a molecule expressed as a ratio of the mass of the molecule to the mass of 1/12 of a ^{12}C atom. Since 1/12 of a ^{12}C atom has a mass of 1 D, the magnitude of the M_r is the same as the magnitude of the MW. So, why bother ? Because M_r is a ratio, it is dimensionless, which simplifies comparisons. In practice, because the magnitude of the number is the same as the molecular weight, there can be confusion. If you are reporting the molecular weight of a protein, you should write MW = 10 kD or 10,000 D and refer to it as the 10 kD protein. If you are reporting the relative molecular weight, you should write M_r = 10,000 (no units because M_r is dimensionless) and refer to it as the 10,000-M_r protein.

THE PERIODIC TABLE

The periodic table shows the relationships among the known (and some unknown) elements. If you read from left to right in the periodic table, you will read the names of the elements in order of increasing atomic number and mass. If you read down a column in the periodic table, you will read the names of elements that have the same number of electrons in their outermost shell and that, therefore, have similar chemical properties.

In the periodic table on the facing page, shading emphasizes those elements that are particularly prevalent in living systems. Hydrogen, carbon, nitrogen, oxygen, phosphorus, and sulfur account for approximately 98% of the weight of life. Often, in color publications and in ball-and-stick diagrams, these elements are color coded according to an accepted convention: hydrogen is white; carbon is gray; nitrogen is blue; oxygen is red; and phosphorus and sulfur are yellow.

This periodic table, like most, contains a number of pieces of information about each element. The symbol for the element is the two-letter abbreviation in the center of each box; the full name of the element is written out below it. In the upper left-hand corner of the box is the atomic number of the element. The atomic number is the number of protons in the nucleus of one atom, and it is the number of protons that determines the element. In the upper right-hand corner is the atomic mass, or atomic weight, of the element. The atomic mass is the mass contributed by all the protons and neutrons and electrons (although the mass contributed by an electron is negligible compared to the combined mass of a proton and neutron). If there is more than one isotope of an element (i.e., if there are versions of that element with the same number of protons but different numbers of neutrons), the atomic mass listed is the mass of the most common naturally occurring form (see table on facing page), or some kind of weighted average of the masses of all isotopes of that element. The organization of the periodic table itself also provides information; i.e., the column and row an element is listed in tells something about the element.

Rows Are Called Periods: There are seven periods. Elements in the same period have the same number of energy levels or valences. For example, Period 1 includes hydrogen and helium. They each have electrons in a single valence shell; the first valance shell is full when it contains two electrons, as it does in helium. Period 2 includes lithium through neon. These eight elements each have electrons in two valence shells; the second valence shell is full when it contains eight electrons. Neon, therefore, has ten electrons in its resting state, two in the first shell and eight in the second.

Columns Are Called Groups: There are 18 groups. Elements in the same group have the same number of electrons in their outermost energy level; so they have very similar chemical properties, such as how they combine with other elements. For example, hydrogen, lithium, sodium, and potassium are four of the members of Group 1. They can each combine with a single other atom such as chloride ion (HCl, LiCl, NaCl, KCl) because they each have one electron in their outermost valence shell.

Legend:

- Group number — 14
- Atomic Number (Number of protons) — 6
- Atomic weight — 12.01
- Symbol — C
- Name — CARBON

Periodic Table

1	2	3	4	5	6	7	8	9	10	11	12	13	14	15	16	17	18
1 1.008 **H** HYDROGEN																	2 4.003 **He** HELIUM
3 6.941 **Li** LITHIUM	4 9.012 **Be** BERYLLIUM											5 10.81 **B** BORON	6 12.01 **C** CARBON	7 14.01 **N** NITROGEN	8 16.00 **O** OXYGEN	9 19.00 **F** FLUORINE	10 20.18 **Ne** NEON
11 22.99 **Na** SODIUM	12 24.31 **Mg** MAGNESIUM											13 26.98 **Al** ALUMINIUM	14 28.09 **Si** SILICON	15 30.97 **P** PHOSPHORUS	16 32.07 **S** SULFUR	17 35.45 **Cl** CHLORINE	18 39.95 **Ar** ARGON
19 39.10 **K** POTASSIUM	20 40.08 **Ca** CALCIUM	21 44.96 **Sc** SCANDIUM	22 47.87 **Ti** TITANIUM	23 50.94 **V** VANADIUM	24 51.10 **Cr** CHROMIUM	25 54.94 **Mn** MANGANESE	26 55.85 **Fe** IRON	27 58.93 **Co** COBALT	28 58.69 **Ni** NICKEL	29 63.55 **Cu** COPPER	30 65.39 **Zn** ZINC	31 69.72 **Ga** GALLIUM	32 72.61 **Ge** GERMANIUM	33 74.92 **As** ARSENIC	34 78.96 **Se** SELENIUM	35 79.90 **Br** BROMINE	36 83.80 **Kr** KRYPTON
37 85.47 **Rb** RUBIDIUM	38 87.62 **Sr** STRONTIUM	39 88.91 **Y** YTTRIUM	40 91.22 **Zr** ZIRCONIUM	41 92.91 **Nb** NIOBIUM	42 95.94 **Mo** MOLYBDENUM	43 [97.9] **Tc** TECHNETIUM	44 101.1 **Ru** RUTHENIUM	45 102.9 **Rh** RHODIUM	46 106.4 **Pd** PALLADIUM	47 107.9 **Ag** SILVER	48 112.4 **Cd** CADMIUM	49 114.8 **In** INDIUM	50 118.7 **Sn** TIN	51 121.8 **Sb** ANTIMONY	52 127.6 **Te** TELLURIUM	53 126.9 **I** IODINE	54 131.3 **Xe** XENON
55 132.9 **Cs** CESIUM	56 137.3 **Ba** BARIUM	57–71 *	72 178.5 **Hf** HAFNIUM	73 181.0 **Ta** TANTALUM	74 183.8 **W** TUNGSTEN	75 186.2 **Re** RHENIUM	76 190.2 **Os** OSMIUM	77 192.2 **Ir** IRIDIUM	78 195.1 **Pt** PLATINUM	79 197.0 **Au** GOLD	80 200.6 **Hg** MERCURY	81 204.4 **Tl** THALLIUM	82 207.2 **Pb** LEAD	83 209.0 **Bi** BISMUTH	84 [209] **Po** POLONIUM	85 [210] **At** ASTATINE	86 [222] **Rn** RADON
87 [223] **Fr** FRANCIUM	88 [226] **Ra** RADIUM	89–103 **	104 [263] **Rf** RUTHERFORDIUM	105 [262] **Db** DUBNIUM	106 [266] **Sg** SEABORGIUM	107 [264] **Bh** BOHRIUM	108 [269] **Hs** HASSIUM	109 [268] **Mt** MEITNERIUM	110 [281] **Uun** UNUNNILIUM	111 [272] **Uuu** UNUNUNIUM	112 [277] **Uub** UNUNBIUM	113	114 [289] **Uuq** UNUNQUADIUM	115	116	117	118

* Lanthanides:

57 138.9 **La** LANTHANUM	58 140.1 **Ce** CERIUM	59 140.9 **Pr** PRASEODYMIUM	60 144.2 **Nd** NEODYMIUM	61 **Pm** PROMETHIUM	62 150.4 **Sm** SAMARIUM	63 152.0 **Eu** EUROPIUM	64 157.3 **Gd** GADOLINIUM	65 158.9 **Tb** TERBIUM	66 162.5 **Dy** DYSPROSIUM	67 164.9 **Ho** HOLMIUM	68 167.3 **Er** ERBIUM	69 168.9 **Tm** THULIUM	70 173.0 **Yb** YTTERBIUM	71 175.0 **Lu** LUTETIUM

** Actinides:

89 [227] **Ac** ACTINIUM	90 232.0 **Th** THORIUM	91 231.0 **Pa** PROTACTINIUM	92 238.0 **U** URANIUM	93 [237] **Np** NEPTUNIUM	94 [244] **Pu** PLUTONIUM	95 [243] **Am** AMERICIUM	96 [243] **Cm** CURIUM	97 [247] **Bk** BERKELIUM	98 [247] **Cf** CALIFORNIUM	99 [251] **Es** EINSTEINIUM	100 [252] **Fm** FERMIUM	101 [257] **Md** MENDELEEVIUM	102 [258] **No** NOBELIUM	103 [259] **Lr** LAWRENCIUM

- *Group 1* elements (not including hydrogen) are called the alkali metals. They are highly reactive.

- *Group 2* elements are called the alkaline earth metals. They are reactive, but not as reactive as Group 1.

- *Group 17* elements are called halogens. They are highly reactive.

- *Group 18* elements are called noble gases. They are very stable.

- *Groups 1, 2, and 13–18* elements are called the main group elements.

- *Groups 3–12* elements are called the transition elements.

- *La-Lu (the lanthanide series) and Ac-Lr (the actinide series)* are called the inner transition elements.

Plutonium is the largest naturally occurring element. Elements larger than plutonium exist, or have existed for an infinitesimally short amount of time, as a result of scientists working with particle accelerators. The elements with the prefix Unun- have been synthesized and detected, and they await official names. Elements 113 and 115–118 are expected, but not yet detected. The origins of the less obvious symbols in the Periodic Table are as follows:

Element	Symbol	Origin
Antimony	Sb	Stibium
Copper	Cu	Cuprum
Gold	Au	Aurum
Iron	Fe	Ferrum
Lead	Pb	Plumbum
Mercury	Hg	Hydrargyrum ("liquid silver")
Potassium	K	Kalium
Silver	Ag	Argentum
Sodium	Na	Natrium
Tin	Sn	Stannum
Tungsten	W	Wolfram

IONS

An ion is an atom or molecule that has acquired a net electric charge by gaining or losing electrons. For example, if calcium loses the two electrons in its outermost energy level, it becomes the ion Ca^{2+}. Many ions have important functions in biological systems. For example, the transmission of impulses along a neuron depends on the movement of sodium and potassium ions (Na^+ and K^+) across the neuronal membrane. Ion balance and water balance, important aspects of physiological homeostasis, are intimately related because water follows ions into and out of cells.

The Nernst and Goldman Equations

The Nernst and Goldman equations describe the relationship between the concentration of ions on opposite sides of a semipermeable membrane and the electrical potential across that membrane.

The Nernst equation

Formulated by Hermann Walther Nernst in 1888, this equation states that when there are different concentrations of a particular ion on different sides of a membrane which is permeable to that ion due to channels through that membrane (i.e., there is a concentration gradient across a selectively permeable membrane), there will be an electrical potential across that membrane due to that ion. Essentially, this means that the ions, if they move down their gradient through the channels, can do work. That deceptively simple observation explains a lot of biological functions. For example, there is an Na^+ gradient across the membrane of most cells, including the cells that line the small intestine. Those cells use the energy stored in that gradient to move glucose out of the gut and into the body, an important function when you are hungry.

The Nernst equation tells you the value of the equilibrium potential (E), in volts, for a particular ion. E is determined by the valence (z) of the ion under consideration (which affects the ability of the ion to cross the membrane), the concentration of the ion on the side of lower concentration ($[X_I]$), and the concentration of the ion on the side of higher concentration ($[X_{II}]$). Also relevant are R, the universal gas constant; T, the temperature in kelvins; and F, the Faraday constant. The Nernst equation:

$$E = \frac{RT}{Fz} \ln \frac{[X_I]}{[X_{II}]}$$

can be written:

$$E = 8.617 \times 10^{-5} \frac{T}{z} \ln \frac{\text{Lower concentration}}{\text{Higher concentration}}$$

Other versions of the equation can be useful, depending on the application. In the version above:

E = equilibrium potential for the ion in question [V]
R = universal gas constant [8.314 J mole^{-1} K^{-1}]
T = temperature [K]
F = Faraday constant [9.649 × 10^4 C mole^{-1}]
z = valence of the ion
ln = "natural log of"
$[X_I]$ = concentration of the ion on side I of the membrane [M]
$[X_{II}]$ = concentration of the ion on side II of the membrane [M]
Lower concentration = concentration of ion on the side with lower concentration
Higher concentration = concentration of ion on the side with higher concentration

E is negative, because the ln of a number less than one is negative. A negative E means the ions could (potentially) flow down their gradient.

The Goldman equation (also known as the Hodgkin-Katz-Goldman equation)

In practice, there is typically more than one kind of ion hanging around the membrane. If you are interested in the conduction of nerve impulses, for example, you are interested in, at least, K$^+$, Na$^+$, Cl$^-$, and possibly Ca^{2+}. The Goldman equation is a version of the Nernst equation that allows you to account for the contributions of many ions to a gradient, not just one species. In the Goldman equation, rather than account for the various valences of the many ions, the permeability of the membrane to each ion is accounted for separately by the terms P_C (permeability of the cation species) or P_A (permeability of the anion species). The Goldman equation is used to calculate the membrane potential, V_m, for biological membranes:

$$V_m = \frac{RT}{F} \ln \frac{\Sigma P_C[C^+]_o + \Sigma P_A[A^-]_i}{\Sigma P_C[C^+]_i + \Sigma P_A[A^-]_o}$$

R = universal gas constant [8.314 J mole^{-1} K^{-1}]
T = temperature [K]
F = Faraday constant [9.649 × 10^4 C mole^{-1}]
P_C = permeability coefficient for the cation [m s^{-1}]
P_A = permeability coefficient for the anion [m s^{-1}]
V_m = membrane potential [V]
$[C^+]_i$ = concentration of any cation inside the cell [M]
$[C^+]_o$ = concentration of any cation outside the cell [M]
$[A^-]_i$ = concentration of any anion inside the cell [M]
$[A^-]_o$ = concentration of any anion outside the cell [M]

Note: For cations, the extracellular (outside) concentration is in the numerator; for anions, the extracellular concentration is in the denominator.

Frequently, for biological membranes, this equation takes the form:

$$V_m = \frac{RT}{F} \ln \frac{P_K[K^+]_o + P_{Na}[Na^+]_o + P_{Cl}[Cl^-]_i}{P_K[K^+]_i + P_{Na}[Na^+]_i + P_{Cl}[Cl^-]_o}$$

which can be written as:

$$V_m = 8.617 \times 10^{-5}T \ln \frac{P_K[K^+]_o + P_{Na}[Na^+]_o + P_{Cl}[Cl^-]_i}{P_K[K^+]_i + P_{Na}[Na^+]_i + P_{Cl}[Cl^-]_o}$$

Note: Other ions, notably H^+, can contribute to V_m.

The permeability coefficients of some biologically important ions in mammals are shown below:

Ion	Permeability Coefficient (P)
K^+	5×10^{-9} m s^{-1}
Na^+	5×10^{-11} m s^{-1}
Cl^-	1×10^{-10} m s^{-1}

Ions and Bioelectricity

Ions are everywhere in organisms and in their environments. When ions move, that is a current; so, electricity is also everywhere. It is actually called bioelectricity in the context of living organisms, and prominent subjects of study are bioelectricity in the nervous system and in the heart. Electricity is another one of those characteristics that is spoken of largely by the numbers.

Quantifiable characteristics of bioelectricity include voltage, for example, the resting potential of a membrane, which can affect the activity of proteins in that membrane, and capacitance, for example, the capacitance of a plasma membrane, which is proportional to its area. Although electricity is difficult to understand on a profound level, it is fairly easy to measure and to describe using equations, because all of the interesting characteristics of electricity are interrelated. Below is a list of electrical phenomena and the relationships among them.

Capacitance: The measure of energy stored by a capacitor. A capacitor comprises two conductors separated by a nonconducting (insulating) substance. For example, a lipid bilayer is a capacitor. If there is a voltage across the capacitor, positive charge accumulates in one plate (one side of the membrane) and negative charge accumulates in the other (nothing actually moves between the plates because of the insulation). The symbol for capacitance is C, its dimensions are $M^{-1}L^{-2}T^4A^2$; it is measured in farads [F]. One farad is the energy-storing capacity of a capacitor that accumulates one coulomb of charge on each plate as a result of a one voltage potential difference across the plates. In other words, one farad equals one coulomb per volt.

Charge (Electrical Charge): This comes in two flavors: positive and negative, the names given to the two "versions" of electricity. Ions have charges, as do polar molecules; the charge on amino acid side chains is a critical determinant of the shapes of proteins. Electrical charge, Q, has the dimensions TA (so it is a duration of current), and it is measured in coulombs (C). One coulomb is roughly the amount of charge in 1/96,500th of a gram of electrons. A single electron has a charge of -1.6×10^{-19} C.

Current: Current is the velocity of a charged species. Velocity is a vector and thus it has both magnitude and direction. The direction of a current is, by convention, the direction of movement of a positively charged species. (This can be a little confusing: if Na^+ and Cl^- ions are physically moving in the same direction, their respective currents are moving in opposite directions.) There are many, many, many ion currents in cells, and they keep us alive (ion currents in mitochondria and the heart), moving (ion currents in muscle cells), thinking and sensing (ion currents in neurons), reproducing (ion currents in eggs), and just about everything else. Current, I, has the fundamental dimension A, and it is measured in amperes (A). One ampere is one coulomb of charge per second.

Resistance: Resistance inhibits current flow. Resistance, R, has dimensions $ML^2T^{-3}A^{-2}$ (or the same dimensions as voltage per ampere) and it is measured in ohms (Ω). A length of conductor that has a resistance of one ohm will allow one ampere of current to flow when there is a potential of one volt across that length. If the resistance were higher, less current would flow.

Voltage: Voltage is a force, specifically, the electromotive force (emf); that is, the force that moves charged things from a point of lower potential to a point of higher potential. It is also called the potential, as in "There is a potential of -60 mV across that membrane." Voltage, V, has dimensions $ML^2T^{-3}A^{-1}$ (it is probably more useful to think of it as the work per charge) and it is measured in volts (V). A one-volt potential means it would take one joule of work to move one coulomb of charge from the position of lowest potential to the position of highest potential. A -60 mV potential across a plasma membrane means it would take 60 mJ to move one coulomb of positive charge across that membrane and out of the cell.

The relationships among these phenomena are described by the equations below.

Ohm's law

The relationship among current, voltage, and resistance is described by Ohm's law:

$V = IR$

V = voltage in volts [V]
I = current in amperes [A]
R = resistance in ohms [Ω]

The relationship can also be written as:

$I = V/R$ for current in amperes
$V = IR$ for voltage in volts
$R = V/I$ for resistance in ohms

When ion channels in cell membranes are discussed, these symbols can be used specifically to mean:

I = current through single channel
V = voltage across channel
R = resistance of open channel

Capacitance

Capacitance, which can be used to measure surface area of a membrane, is given by:

C = capacitance in farads [F] = Q/V

Q = charge in coulombs [C]
V = voltage or potential difference across the capacitor, in volts [V]

Voltages, resistances, and capacitances can be added; how to calculate the sum depends on whether they are connected in series (end to end) or in parallel (side by side).

Adding voltages of potentials connected in series:

$V = V_1 + V_2 + V_3 + ... + V_n$

Adding resistance of resistors connected in series:

$R = R_1 + R_2 + R_3 + ... + R_n$

Adding resistance of resistors connected in parallel:

$1/R = 1/R_1 + 1/R_2 + 1/R_3 + ... + 1/R_n$

Adding capacitance of capacitors connected in series:

$1/C = 1/C_1 + 1/C_2 + ... + 1/C_n$

Adding capacitance of capacitors connected in parallel:

$C = C_1 + C_2 + C_3 + ... + C_n$

Quantifying Chemical Reactions

Chemical reactions can be written as expressions, which, although they are not mathematical equations, are referred to as equations. What chemical equations have in common with mathematical equations is that the left-hand side and right-hand side have to balance. In the case of chemical equations, this means that the number of atoms of each element on the left-hand side must match the number of atoms of that element on the right-hand side. A first check of whether a chemical equation is correct is to count the number of atoms going into a reaction and compare that with the number coming out in the products. If the numbers don't match, something is wrong; remember Lavoisier's law of conservation of mass: matter is neither created nor destroyed. If an atom goes into a chemical equation as a reactant, it must appear in the products. Here, confusion can arise because sometimes reactants or products, such as H_2O, are left out. Below is the equation describing the chemical reaction that makes your life on earth possible:

$$6\ CO_2 + 6\ H_2O \rightarrow C_6H_{12}O_6 + 6\ O_2$$

This equation shows the net change that results from photosynthesis. To check that this equation balances, count the number of Cs, Hs, and Os on the left-hand side and confirm that it matches the number on the right-hand side.

Left: $6 \times 1 = 6$ carbons; $6 \times 2 = 12$ hydrogens; $6 \times 2 + 6 \times 1 = 18$ oxygens
Right: $1 \times 6 = 6$ carbons; $1 \times 12 = 12$ hydrogens; $1 \times 6 + 6 \times 2 = 18$ oxygens

It is balanced.

The Equilibrium Constant

The equilibrium constant is an important number that is used to characterize reactions. When a reaction is at equilibrium, i.e., no net change in the concentration of reactants or products, there is a constant relationship between the concentrations of reactants and products. This makes sense if you realize that when you add reactants to a reaction at equilibrium, you will quickly generate more products, and vice versa. (At equilibrium, the reaction is going both directions. By convention, reactants are the chemicals on the left-hand side of the equation as written and products are the chemicals on the right-hand side of the equation as written.) The relationship between the concentrations of products and reactants in a reaction at equilibrium can be summed up in a single number called K_{eq}. K_{eq} for a given reaction (at a given temperature) is a constant, hence, the equilibrium constant.

K_{eq} is a ratio: The numerator is the product of the concentrations of the products, each raised to a power equal to its coefficient; the denominator is the product of the concentrations of the reactants, each raised to a power equal to its coefficient. This is much easier to show than to describe; here is an example:

A reaction at equilibrium: $N_2 + 3 H_2 \rightleftarrows 2 NH_3$

For this reaction: $K_{eq} = [NH_3]^2/[N_2][H_2]^3$

Reaction Rate

Another important characteristic of a reaction is how fast it takes place, i.e., the reaction rate. The reaction rate is the rate at which the reactants become products. Products must be appearing at the same rate that reactants are disappearing (otherwise matter would be disappearing into thin air or appearing out of thin air). If you are interested in a reaction rate, you can measure *either* the change in concentration of reactants (which will be disappearing at the reaction rate) *or* the change in concentration of products (which will be appearing at the reaction rate). In practice, it is most appropriate to measure the concentration of the product. Reaction rates are usually measured as concentration of product appearing per unit time, thus, the units of a reaction rate can be moles per liter per second [mole l^{-1} s^{-1}]; but they can also be [g s^{-1}] or [mole s^{-1}] or any other way you would like to measure the change in amount of product over time.

Like any other rate, a reaction rate is determined by graphing time on the x axis, then molarity or amount on the y axis, and then taking the slope. The slope of the curve describing the data is the reaction rate.

Concentration and Reaction Rate: The Rate Law, Rate Constants, and Reaction Order

The rate of a reaction can be affected by the concentration of reactants and the temperature. The effects of concentrations are summed up numerically as the rate constant and the equilibrium constant.

Take some generic reaction, 2A + B → C. The rate of this reaction is the change in the concentration of C over time. That can be rewritten symbolically as $d[C]/dt$ (*d* means "the change in"; $[C]$ means "the concentration of C," *t* means "time").

It also makes intuitive sense that if you have more A and more B, the reaction will go faster. Put yourself in the place of an A that has to bump into a B and another A to react and make a C. If there is a higher concentration of A and B around, you will bump into one sooner. That will be true of all the As and all the Bs, who are all bumping into each other sooner; so the entire process goes faster. The rate of the reaction, $d[C]/dt$ is therefore a function of the concentration of the reactants $[A]$ and $[B]$. Put symbolically, $d[C]/dt = f([A],[B])$. The function, generically speaking, is the product of the concentrations of the reactants, with each raised to the power of its coefficient. So, for this particular reaction (2A + B → C), the function looks like this:

$$d[C]/dt = k[A]^2[B]$$

This is the *rate law*, or rate equation, for this reaction (reactions that look different will have different looking rate equations). *k*, the proportionality constant that tells you the relationship between the reaction rate and the concentration of the reactants, is called the *rate constant*.

Another number used to characterize reactions is the *reaction order*. The order of a reaction is calculated by adding up the exponents on the concentration terms in the rate equation. If the reaction looks like A + B + 2C → D, the rate equation looks like $d[D]/dt = k[A][B][C]^2$. The sum of the exponents is $1 + 1 + 2 = 4$, making it a fourth-order reaction. The reaction order tells you about the dependence of the reaction rate on the concentrations of reactants. In practice, experimentation gives you the exponents of the rate equation. For this particular reaction, doubling the concentration of A or B would double the rate of the reaction, whereas doubling the concentration of C would quadruple the rate of the reaction. From these data, you can write the rate equation; from the rate equation you can write the expression for the reaction.

The effect of temperature on reaction rate: The Arrhenius equation

When studying reactions, the effect of temperature on the rate of the reaction is an important consideration. For example, if you are studying an enzyme-catalyzed reaction, the graph of rate versus temperature will increase up to a point, and then decrease again, because above a certain temperature, the enzyme will become denatured and will not work as well. The highest point on the curve gives the optimal temperature for the enzyme.

To see if the reaction rate that you are interested in varies with temperature, you measure the rate at various temperatures. First, you measure concentration of product at different times, then take the slope of the graph of concentration versus time, then take those same measurements again at a different temperature, then again at a third temperature, and so on. Say you measure concentration of product over time at 16 different temperatures. At the end of all that measuring, you will have to make 16 different graphs, each representing how concentration (*y* axis) changed with time (*x* axis) at one given temperature.

But you aren't done yet. You now want to know if there is a relationship between reaction rate and temperature. To go from 16 separate graphs of concentration versus time to one graph of reaction rate versus temperature, you take the slopes (rates) from the 16 graphs and make a new graph with the rates (slopes) on the *y* axis and the temperature on the *x* axis. Graph your 16 points. Now, you look at this new graph of rate versus temperature, and you will be able to see if there is a relationship between temperature and rate.

Generally speaking, for simple reactions, the higher the temperature, the faster the reaction rate. This relationship is intuitive when you consider that reactants need to collide with sufficient energy to react: If it is hotter, the molecules will be moving around faster and are more likely to collide with sufficient energy to react. The relationship between temperature and reaction rate is described by the Arrhenius equation:

$$k = A \exp(-E_a/RT) = Ae^{-E_a/RT}$$

k = rate constant of the reaction
A = a constant
e = base of the natural log, i.e., 2.72
E_a = activation energy [J mole^{-1}]
R = universal gas constant [8.3145 J mole^{-1} K^{-1}]
T = temperature [K]

A calculator at www.shodor.org/unchem/advanced/kin/arrhenius.html will calculate the value of any of the variables in the Arrhenius equation, given the other three.

Summing up the effect of temperature on reaction rate: Van't Hoff and the Q_{10}

It is common to see the effect of temperature reported as the ratio of reaction rates measured at two different temperatures. A difference of 10°C has become a kind of standard for measuring, and the value of the ratio is given by the Van't Hoff equation, which defines Q_{10}. Q_{10} is a dimensionless number (because it is a ratio) that sums up the effect of temperature on the rate of a reaction:

$$Q_{10} = \left(\frac{k_2}{k_1}\right)^{10/(T_2 - T_1)}$$

where T_2 and T_1 are the two temperatures at which the rate was measured and k_2 and k_1 are the rate constants at those two temperatures. Q_{10} can also be calculated by the equation:

$$Q_{10} = \left(\frac{U_2}{U_1}\right)^{10/(T_2 - T_1)}$$

where U_2 and U_1 are the reaction rates at T_2 and T_1. If the difference between T_2 and T_1 is exactly 10°C, Q_{10} is given by the very simple equation:

$$Q_{10} = \frac{U_{(T+10)}}{U_T}$$

U_T = the rate at some temperature T [s^{-1}]
U_{T+10} = the rate at 10°C above T [s^{-1}]

☑ Example

If a reaction rate at 26°C is 20 moles per liter per second, and the reaction rate at 36°C is 50 moles per liter per second, the Q_{10} of the reaction is:

$$Q_{10} = \frac{50 \text{ moles/liter/second}}{20 \text{ moles/liter/second}} = 2.5$$

IMPORTANT: Because the difference in reaction rates from 10°C to 20°C may be different from the difference in reaction rates from 20°C to 30°C, it is important to report T_2 and T_1 when reporting a Q_{10}.

RADIOACTIVITY

Radioactivity is the emission of energy due specifically to the decay of an atomic nucleus. The energy is emitted in one or more of the following forms: alpha (α) particles, beta (β) particles, and gamma (γ) rays. Note that X-rays have properties similar to those of γ-rays, but they originate outside the nucleus of an atom. The following table sums up the changes in an atom when its nucleus decays and emits energy in these forms:

Form of Energy	Composition	Change in		
		Atomic Number	Atomic Mass (amu)	Charge
α particle	2 protons + 2 neutrons (equivalent to a helium nucleus)	−2	−4	−2
β particle	electron (emitted only when a neutron decays into a proton and an electron)	+1	negligible	+1
γ ray	radiation of $\lambda < 10^{-11}$ m	−	−	−

Radioactivity is ionizing radiation, i.e., the radiation emitted during decay is energetic enough to displace electrons (hence making ions) from a substance with which it interacts. Types of radiation that do not have sufficient energy to displace electrons, such as microwaves and light, are called nonionizing radiation. This high-energy ionizing radiation is what makes radioactivity dangerous. There are different ways to quantify radioactive decay and its effects. Which quantification is used depends on what you are interested in knowing.

Units for Measuring Radiation

Becquerel (and Curie): This unit tells you the rate of decay of atomic nuclei in disintegrations per second (1 Bq = 1 dps). Becquerels can therefore be used to characterize the quantity of radioactivity over time. Becquerels have dimensions of inverse time (T^{-1}).

Gray: The unit for RAD (abbreviation for radiation absorbed dose). This unit tells you the size of a dose of radioactivity. Grays are the energy absorbed per kilogram of material absorbing; they can describe the dose of any kind of radiation in any material.

Roentgens: A measure of the ionization of air molecules. Roentgens tell you the amount of exposure in seconds of current per kilogram of dry air (coulombs/kg). Roentgens are easy to measure, but they only measure the effect of ionizing radiation (X-rays and γ-rays), and only in air.

Sievert: The unit for REM (abbreviation for roentgen equivalent man). This unit tells you the effect, in energy per kilogram, of radiation on human tissue. REMs are calculated by multiplying the RAD times a coefficient (Q) that is specific to the type of radiation.

The following tables show the units used in measurements of radioactivity.

SI Unit	Symbol	Measure of	Value of 1 Unit	Dimensions
Becquerel	Bq	Rate of decay	1 dps	T^{-1}
Gray	Gy	Absorbed dose	1 J kg^{-1}	L^2T^{-2}
Sievert	Sv	Effective or equivalent dose	1 J kg^{-1}	L^2T^{-2}
Roentgen	R	Exposure to X- and γ-rays	2.5810^{-4} C kg^{-1} of air	T A M^{-1}
Non-SI Unit				
Curie	Ci	Rate of decay	3.710^{10} Bq	T^{-1}
RAD	rad	Absorbed dose	10^{-2} Gy	L^2T^{-2}
REM	rem	Effective or equivalent dose	10^{-2} Sv	L^2T^{-2}

For information about equipment used for measuring radioactivity, see Chapter 3, "Common Laboratory Equipment."

Half-life of a Radionuclide

The half-life of an element, abbreviated $t_{1/2}$, is the amount of time it takes for 50% of that element to disappear, decay, transform, or otherwise become nonfunctional. Different radioactive nuclides (also known as radionuclides or radioisotopes) have different half-lives, meaning they decay at different rates. It takes more than 5000 years for half the amount of ^{14}C in a material to decay, which makes ^{14}C useful for dating certain fossils. Other radionuclides are used as labels for subcellular molecules because the emissions that result from their decay can be imaged on film. The half-life of those radionuclides determines how long the reagents are hazardous; thus, understanding half-lives is an important safety issue.

The total length of time it takes a radionuclide to decay depends on the initial amount of radionuclide present (the same way that the rate of decay of beer foam depends on the amount of foam present). Thus, the relationship between the number of half-lives that have passed and the amount of radionuclide remaining is not linear. The relationship looks like this:

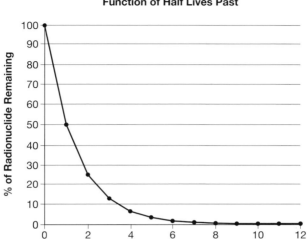

% Radionuclide Remaining as a Function of Half Lives Past

If, over the course of a half-life, only half of the amount of element disappears, theoretically, the remaining amount can never reach zero. Actually, there will come a point where there is one atom or molecule left, and it either disappears or it doesn't. The rule of thumb is that after the passage of ten half-lives, when there is approximately 1/1000th of the original amount left, you will not be able to detect any radiation above background and can consider the radioactivity gone; the limit of your ability to detect remaining radionuclides may be reached sooner. See the Resources list for more information about measuring radioactive decay.

Radionuclides Commonly Used in the Laboratory

Note: Consult radiation safety manuals and local institutional regulations for complete information on the use of radionuclides.

^{32}P Half-life: 14.262 days
Mode of decay: β emission
Decay energy: 1.711 MeV

^{35}S Half-life: 87.51 days
Mode of decay: β emission
Decay energy: 0.167 MeV

^{125}I Half-life: 59.408 days
   Mode of decay: γ- and X-ray emission
   Decay energy: 0.186 MeV

^{3}H Half-life: 12.4 years
   Mode of decay: β emission
   Decay energy: 0.019 MeV

^{14}C Half-life: 5730 years
   Mode of decay: β emission
   Decay energy: 0.156 MeV

The information covered in this chapter is useful background for understanding what you are doing when you work with numbers associated with chemicals and their behavior. In a biology laboratory, a very common example of working with numbers associated with chemicals is the making of solutions. Solution making is another topic characterized by a lot of numbers and a *lot* of confusion. If one of your goals is to make solutions quickly and accurately, read on. Chapter 3 and especially Chapter 4 were written to help with exactly that.

RESOURCES

Arrhenius equation variables calculation

http://www.shodor.org/unchem/advanced/kin/arrhenius.html

Measuring radioactive decay

http://www.curvefit.com/radioactivity_theory.htm

http://www.curvefit.com/calculator.htm

Periodic table

http://pearl1.lanl.gov/periodic/default.htm

http://www.webelements.com

Radiation and radiation safety

http://www.umich.edu/~radinfo/

Equipment for Measuring, Counting, and Otherwise Quantifying

IMPORTANT CONSIDERATIONS WHEN MEASURING

Some very important considerations must be taken into account when any piece of laboratory equipment is used. These are discussed below.

Uncertainty

As far as the laws of mathematics refer to reality, they are not certain; and as far as they are certain, they do not refer to reality.

Albert Einstein (1879–1955)

Life is uncertain; measurements are uncertain. The best thing is to let the reader of your publication know how uncertain you are about any of your measurements. This is not only honest, it is absolutely required in science, and it is how you let others know that any uncertainty associated with the measurements is due to the limitations of your machine, not you. Everyone knows that instruments are not perfect, so it is expected that you will provide some indication of how imperfect the reported value might be. To provide this information, you must know two things about your measurement: its accuracy and its precision. As discussed in Chapter 1, both of these have an effect on the uncertainty of your measurement; so you must know about both the precision and the accuracy of your measuring device.

The precision of a device is measured by how close together repeated measurements are. Precision is often measured and reported by the manufacturer in the instruction manual that comes with the instrument; look for a graph or a number that indicates the range of values associated with repeated measures. If you cannot find any information, just measure the same thing 30 times and calculate the range of values.

The accuracy of a device is measured by how close the measurement is to the actual value. Accuracy is determined for a device by measuring something whose value is known a priori (often called a standard), and then comparing the measured value to the known value. (*Note:* Devices can give measurements that are very precise and completely inaccurate.)

More about Precision

Precision is an evaluation of the consistency of a measured value across repeated measurements. To determine precision, you have to repeat your measurement and determine the range of outputs. To report precision, use the correct number of significant digits; the precision is assumed to be ±1 in the last significant digit (see Chapter 1). Alternatively, you can report precision more explicitly by indicating the range of your measurements exactly, for example, 3.026 ± 0.007 kg. If you use either of these formats, you are reporting the absolute precision. If you report the range as a percentage of the measure (e.g., 420 kg ± 10%), you are reporting relative precision.

More about Accuracy

Accuracy is a measure of truth, or the nearness of your measurement to the correct value. To improve the accuracy of your measurements, calibrate your equipment by measuring standards of known values. Once the device gives the correct measure of a standard of known value, then you have to trust that it will give you a measure of equal accuracy when you are evaluating something unknown. To evaluate and report accuracy, you report a statistic (usually the standard deviation and the *p*-value) that indicates the probability of the measurement being within a certain distance of the true value. For more on this, see Chapter 7. The single most important thing you can do to improve the accuracy of your measurements is to calibrate your equipment.

Calibration

Calibration is the process of testing and adjusting a measuring device to determine the relationship between its output and the true value of the variable being measured. Calibration is accomplished by measuring independently supplied standards; the actual values of the standards are known a priori. In some cases, for example, when calibrating a pH meter, you adjust the device itself, according to the standard, so that the readout is accurate. In other cases, for example, when calibrating a spectrophotometer, you determine the relationship of the device's output to true values by graphing output as a function of the value of the standard. The graph itself, or the equation of the resulting curve, is used to convert measurements of unknowns to their actual values. Such a graph is sometimes called a standard curve.

It is important when calibrating a device to use a range of standards that will bracket the values you expect to measure. For example, if you will be measuring the pH of biological reagents, the values you will be measuring will be around 7; thus, you should use a pH 4 standard and a pH 10 standard to calibrate the pH meter. Similarly, to calibrate a spectrophotometer you should measure standards that bracket the expected concentrations of your unknowns.

It is also important when calibrating certain devices to use standards that are as similar as possible to whatever you will actually be measuring. This is particularly important when calibrating a spectrophotometer. For example, if you will be using it to measure the concentration of a protein in solution, you want the amino acid content of your protein standard to be as similar as possible to that of your unknown, because different amino acids absorb light differently.

For some instruments, an important part of calibrating is telling the instrument what value to call "zero." This process is called taring or zeroing. Spectrophotometers, for example, must be zeroed before every reading; balances must be tared before each weighing. In both cases, you are telling the machine how much of the signal is not relevant. For a balance, you are telling it to subtract the mass of the weigh boat; for the spectrophotometer, you are telling it to subtract the absorption by the medium and the cuvette. It is by taring or zeroing that you make sure that only the value of the unkown is reported, not the value of the measuring container plus the unknown.

Choosing a Measuring Device

Sometimes, you will have a choice between different measuring devices. For example, you probably have many graduated cylinders with a variety of capacities. So, which do you use? You should always choose the size of any measuring device based on how much you need to measure: The amount to be measured should, whenever feasible, fall in the middle of the range of measurements possible with the instrument. Another good rule of thumb is that the amount of material to be measured should never be less than 10% of the capacity of the instrument.

MEASURING LIQUIDS (VOLUME)

Although beakers and flasks are typically marked on the side with volume amounts, these pieces of laboratory equipment are not accurate measuring tools and should only be used as mixing vessels.

Graduated Cylinders

Graduated cylinders are used to measure volumes of fluid between 1 ml and 2000 ml. To use a graduated cylinder properly, your eye must be level with the surface of the fluid being measured. The lines on most graduated cylinders are designed to help you line up your eye correctly. From at least one angle, you can see each line both directly *and* through the cylinder. Your eye is at the correct level to evaluate the volume in the cylinder when the two halves of the line are superimposed, i.e., when the line you can see directly exactly covers the line on the other side of the cylinder that is visible through the glass or plastic. If you can only see the line from one angle, do your best to put your eye level with the line.

METHOD How to Use a Graduated Cylinder

1. Determine which line on the cylinder represents the volume you wish to measure.

2. Add fluid to the cylinder until it gets somewhat close to the line.

3. Put your eye level with the line, as described above, and finish adding fluid. When the bottom of the meniscus is level with the line, which is level with your eye, you have the proper volume.

4. If you overshoot, you can remove the fluid with a Pasteur pipette.

Pipettes (Serological Pipettes)

Pipettes are used to measure small volumes of fluid (typically 0.1–25 ml, although you can purchase pipettes that will hold 50 ml or 100 ml). Both the maximum capacity and the size of the increments are printed on the end of the pipette.

There are two kinds of pipettes: TC ("To Contain") and TD ("To Deliver"). There is a very important difference between them: TC pipettes will release the correct volume just by gravity; TD pipettes will not, and so the fluid left in the tip must be "blown out." Most pipettes are TD. The pipette may be marked TC or TD at the end.

Like graduated cylinders, pipettes are often marked with lines that encircle most or all of the pipette to assist you in properly lining up your eye to evaluate your measurement. Some pipettes have ascending gradations, others have descending gradations, and still others have both, one on either side of the pipette. There is no functional difference between the styles of pipettes; you just need to account for the type you are using.

For example, say you are using a 10-ml capacity pipette to measure out 7.0 ml of solution. If the pipette has zero at the tip, you will fill the pipette to the line demarcating 7.0 ml and then dispense the entire contents. If the pipette has zero at the end (meaning the top, the part where you *never* put your mouth), you can fill the pipette to the line demarcating 3.0 ml (10 – 7.0) and then dispense the contents.

METHOD **How to Use a Pipette**

IMPORTANT: *Never* mouth pipette! Even if you are pipetting water, there may be something unseen on the pipette itself that you do not want in your mouth.

1. Attach the pipette filler/dispenser of choice to the end of the pipette. Pipette aids differ, but they all have a way to pull fluid into the pipette and a way to push it out. Many have a separate way to blow out the bit of fluid that gets trapped in the tip of TD pipettes.

2. Submerge the pipette tip a few millimeters into the source. Keep the pipette almost vertical.

3. Use the pipette aid to pull fluid into the pipette until the bottom of the meniscus reaches the desired line.

4. Move the pipette to the receiving container and submerge the tip below the surface of the solution to which you are adding the measured fluid. If the receiving container is empty, touch the pipette tip to the side of the container. This minimizes bubbling.

5. Use the pipette aid to push out the measured fluid. If necessary, blow out the last bit for an accurate volume.

Pipetters

Pipetters are mechanical pipettes used to measure volumes ranging from 0.1 μl to 5000 μl. All pipetters have a plunger with two stops, a barrel, an extension that holds a disposable tip, and a pipette tip ejector; some also have a filter that prevents fluid from entering the barrel. Some pipette tips have their own additional filters that help to prevent contaminants from entering the fluid. Some pipetters are set to pick up one volume, whereas others can hold a range of volumes. Adjustable pipetters have a dial (manual pipetter) or a digital display (electronic pipetter) that allows you to set the desired volume. The maximum capacity of an adjustable pipetter is typically indicated on the top of the plunger. The dial is marked in microliter increments, and the position of the decimal point is usually indicated by a colored line.

METHOD   How to Use a Manual (Nonelectronic) Pipetter

1. Set the volume by turning the dial. Never go above the volume print-ed on the plunger (the maximum); never go below the minimum. A good rule of thumb is not to go below 10% of the maximum.

2. Open a box of disposable pipette tips and push the pipetter extension securely into the tip. To avoid contamination, never allow the tip to touch anything else.

3. Depress the plunger to its first stop.

4. Slightly submerge (about 3 mm of the tip) the tip into the fluid.

5. *Slowly* release the plunger. This pulls the fluid into the pipette tip. If this is done quickly, the tip may not fill properly and, worse, fluid may splash the filter. If the filter gets wet, the pipetter will not work properly; you must change the filter before continuing or your vol-umes will be wrong. If you find that fluid drips out of the pipette tip, it may be that the filter has gotten wet.

6. Put the tip into the destination container. If the container is empty, put the tip near the bottom of the container, or touching the side. If the container already has some fluid in it, submerge the pipette tip about 3 mm.

7. Depress the plunger to its first stop, using slow and even pressure. If necessary, you can, at this point, release the plunger again, slowly, and depress it again, to clean out the inside of the pipette tip. If you do this, do not depress the plunger past the first stop.

8. Finally, depress the plunger to its second stop and hold it there, blow-ing out the remaining liquid in the tip. Keep the plunger depressed as you remove the tip from the fluid.

9. Eject the tip into an appropriate waste container as determined by local institutional safety regulations.

MEASURING DRY CHEMICALS (WEIGHT OR MASS)

Balances

Balances measure weight but report mass. The units printed on balances are often units of mass, such as the SI unit grams. In practice, balances measure weight and then convert weight to a measure of mass by dividing automatically by gravity. (Gravity is different at different heights above sea level, which is one reason why it is important to have your balance calibrated and used in the same place. Depending on its sensitivity, moving a balance to a room two floors up could affect its accuracy.)

If you read the instruction manual for the balance in your laboratory, you will find that there are quite a few phenomena that can affect the values a balance reports. These phenomena include electrostatic charges, humidity, and the height of the balance relative to sea level. Some of these phenomena affect how the balance interprets the forces on it; some change the actual weight of the sample; some affect the value of gravity. There are equations for correcting the measurements to account for the various phenomena, but it is simpler to always follow procedures that minimize distortions. For example, position balances in the laboratory away from the air-conditioning vent, which can cause variation in measurements as the air blows. A balance positioned near a centrifuge can be affected by vibration. Balances must be calibrated in situ and recalibrated if they are moved. Finally, be careful not to underestimate uncertainty in your measurement. The following advice will help ensure that your balance works optimally.

METHOD How to Best Treat a Balance

1. Make sure that the balance is always level. Good balances have bubble levels; the bubble should be in the center of the ring.

2. Calibrate the balance in situ. A balance should always be calibrated in the location where it will actually be used—same room, same bench top, same place on the bench top. If you have professionals do the calibrating, they will do it in situ. You cannot send a balance in for calibrating.

3. Always keep your balance clean.

METHOD How to Best Use a Balance

1. Calibrate the balance using standard weights, the built-in calibration function, or a professional. If your balance is electronic, let it warm up before calibrating or using it.

2. Wear gloves. This is not just a safety issue; it is done to protect what you are weighing from the effects of being near moist, oily skin.

3. Always use a weigh boat or weigh paper, and make sure the pan of the balance and the weigh boat are clean.

4. Zero the balance by pushing the tare button; zero with the weigh boat on the balance pan to account for its weight.

5. With the weigh boat already on the pan and the balance zeroed, put the sample into the weigh boat and read off the answer. The units will be printed next to the readout, and the decimal place will be indicated. If you are measuring a substance to a certain amount, continue to add the substance to the weigh boat until you have the desired quantity. (A trick for adding very small amounts of dry chemical is to hold the scoopula over the weigh boat and, using your other hand, gently tap the arm or hand that is holding the scoopula.) If you overshoot your target amount, remove some of the substance until the appropriate amount remains in the weigh boat. Alternatively, put a new weigh boat on the pan, and start over, using the chemical in the first weigh boat as the source. Do NOT return material back to the original container. This will lead to contamination of your stock.

   Another alternative is to recalculate the amount of solvent you would need for this slightly higher amount of chemical; adjusting fluid volumes can be easier than adjusting amounts of dry chemicals.

> **IMPORTANT:** Do not believe a measurement from a balance unless the display is steady. When the display wobbles, it is an indication that something is wrong. Sometimes an environmental factor, such as vibrations from a centrifuge, is disturbing the balance. To diagnose what is wrong, go to the instructions for your particular balance.

COUNTING AND MEASURING WITH MICROSCOPES

Microscopes are amazing tools. They magnify minuscule things by manipulating the light that is coming from those things. Microscopes can also tweak the light illuminating an object in such a way that otherwise transparent objects become visible to the eye. In addition to the benefits of being able to see a microscopic object, it is often valuable to have some measure of the size of a microscopic object, such as a cell or subcellular organelle. Such information can be useful for identification purposes, as well as for assaying effects of experimental treatments or for making comparisons. Many of the more sophisticated types of microscopes, such as electron and confocal microscopes, have built-in functions that can assess the lengths of objects that are indicated by the user. Compound microscopes can be equipped with a device called an ocular micrometer that allows the user to measure the lengths of objects in the field of view. Unlike with the fancier microscopes, however, there is some calculating involved in calibrating an ocular micrometer (much to the dismay of most introductory microscopy students). To understand the calibration, as well as important aspects of light microscopy, it is useful to understand what a microscope does.

There are two main mathematical considerations that are important for understanding microscopy and doing it well: magnification and resolution. These are discussed below.

Magnification

Magnification is defined as the size of the virtual image (the image you see) relative to the size of the actual object. It is reported as a number followed by the symbol x which means "times." Magnification = magnification by the objective lens x magnification by the ocular lens.

Both the objective lens (the one up close to the object), which collects light coming from the object, and the ocular lens (the eyepiece), which collects light coming from the objective, magnify. An objective lens generates a magnified image that is larger than the specimen by a factor that is written on the objective, for example, 20x. That image, which is already 20 times larger than the specimen, is then magnified by the ocular lens, usually by a factor of 10. The result is the image seen by you or recorded by a camera. It is called the virtual image, and it appears as 20 x 10 = 200x the size of the original.

Resolution and the Rayleigh Criterion

Resolution is defined as the smallest distance (d) by which two objects can be separated and still be seen through a microscope as separate objects. The smaller that separation distance is, the better the resolution; therefore, to improve the resolution of an image, you work for a *decrease* in the value of d.

One question to think about when trying to understand resolution is: Why would the image of two separate objects look like only one? The answer is that diffraction—the tendency of a beam of light to spread out after it passes through a small opening—causes the illusion. For example, there is a small opening in your eye (called the iris) that light passes through to reach your retina. If you have ever been on a long, relatively straight highway at night, you may have noticed that when cars are very far off, they look like they only have a single headlight. This occurs because of diffraction of the light at your iris. The light coming from each of the oncoming headlights "spreads" as it passes through your iris, and the spreading causes the two lights to be perceived as one. As the cars get closer, however, the single spot of light resolves into two headlights. What changes as the cars get closer? The distance between the light source and your eye gets shorter; as a result, the *angle* between the two light beams entering your eye increases and the spreading of the two lights stops overlapping. At this point, your eye is able to discern two lights instead of one. The angle at which the two lights will be *visible* as two lights, the critical angle, was figured out by Rayleigh and is thus known as the Rayleigh criterion.

Three things affect resolution when objects are viewed through a microscope (as opposed to on the highway).

1. The wavelength, λ, of the light collected by the microscope.

2. The refractive index, *n*, of the medium between the coverslip and the objective lens through which the light passes (air, water, glycerin, or oil). The refractive index is a measure of light spreading as it passes through the medium. The higher the diffractive index, the less the light spreads; thus, more light will enter the objective.

3. The amount of available light entering the objective lens. This is characterized as the angle θ, which is half the angle of the cone of light leaving the specimen and entering the objective. θ is a function of the position of the condenser, the diameter of the objective, and the distance between the objective and the object (also called the working distance).

Ernst Abbé figured out the relationship among the various factors listed above and how they affect resolution. The equation for that relationship is known as Abbe's law. (*Historical Note:* Abbe worked with Carl Zeiss to develop oil immersion lenses. The refractive index of the oil was matched to that of the glass, thus minimizing any extraneous spreading of light as it passed from the coverslip into the objective lens.)

Abbe's Law

Abbe's law, version 1:

$$d = \frac{0.612\lambda}{\text{N.A.}}$$

d = Resolution [nm]

λ = Wavelength of light [nm] (by convention, 550 nm is used for light microscopy)

N.A. = Numerical aperture of the objective lens (printed on the objective lens)

Abbe's law, version 2:

$$d = \frac{0.612\lambda}{n \sin \theta}$$

d = Resolution [nm]

λ = Wavelength of light [nm] (by convention, 550 nm is used for light microscopy)

n = Refractive index of the medium touching the objective lens (see table below)

θ = half the angle of the cone of light entering the objective [rad]

☑ Example

To get the best possible resolution from a compound microscope, you must have a really good 100x oil immersion lens, which will have a numerical aperture (N.A.) of about 1.45. If you use the conventional 550 nm to represent the wavelength of the illuminating light, a quick, back-of-the-envelope calculation (0.612 x 550 nm ÷ 1.45) tells you the theoretical best resolution that could be achieved is 232 nm, or about 0.23 μm. The actual resolution will rarely be that good (due to things like chromatic aberration, and the fact that illuminating light contains many wavelengths, not just 550 nm). This means that if you are looking through a microscope, or at a micrograph, you will not be able to distinguish as separate any two things that are closer together than 0.2 μm. You therefore have to be careful when interpreting light micrographs: That which appears to be one thing may, in fact, be two or more things that are close together.

Improving the Resolution

When you first look at a specimen with a microscope, there are likely to be some parameters that need tweaking before you have optimized the resolution of the magnified image. The following guidelines address the possible ways to optimize your image. They assume that you have already established the optimal light path, i.e., Koehler illumination.

Wavelength (λ)

In theory, resolution can be improved by illuminating your specimen with shorter-wavelength light. For example, you could illuminate with blue light, the visible color with the shortest wavelength; however, our eyes don't see blue as well as they see green, and so most objectives are designed with built-in corrections that assume green light will be used. Some microscopes have green filters that can be placed into the light path to take advantage of this characteristic of objective lenses. For the most part, however, compound microscopes use white light. In addition, there is a limit to how short a wavelength you can use. UV light has been tried, for example, and it does improve resolution; but it turns out to be impractical because it requires that you use a quartz objective lens, and quartz distorts the image in other ways. Plus, you need special imaging equipment because human eyes cannot see UV.

Refractive index (n)

The higher the refractive index of the media between the coverslip and the lens, the more light emerging from the specimen is bent toward the objective by the medium. If more light enters the objective, resolution is improved. So, in theory, you could improve resolution by using a medium with a higher refractive index. In practice, any particular objective is made to work only while in contact with air, water, glycerin, or oil. Oil, "glycerin," and water immersion lenses will have "water" or "oil" printed on the side and will only work if they are immersed in the appropriate medium. If no medium is listed, the lens should be used in air only. This means that to improve resolution by increasing n, you have to change objectives; i.e., you cannot just add oil.

Angle (θ)

The size of the cone of light entering the objective is a function of the working distance of the lens (how close the lens is to the coverslip when the image is in focus). This parameter is controlled, in part, by the position of the condenser; establishing Koehler illumination, which puts the condenser in the correct place, is therefore important to optimizing resolution. In theory, moving the objective closer to the object will raise θ and thus improve resolution. In practice, for any particular objective, there is only one position where the image will be in focus. So, as with diffractive index, the only way to improve resolution by increasing θ is to use a different objective.

Numerical aperture (N.A.)

Because both n and θ are characteristics of a particular objective lens and cannot be changed (see above), the two values are combined into one value called the N.A. of the lens. The numerical aperture of a lens is equal to the diffractive index of the medium times the sine of θ (N.A. $= n$ sin θ). The N.A. of an objective lens is printed on the side of the lens. To improve resolution by raising N.A., you have to use a different objective.

IMPORTANT: The N.A. of the condenser, which focuses the light before it enters the specimen, also matters. For this reason, most condenser lenses have an N.A. that is greater than the N.A. of the best objective; thus, in practical terms, it is the N.A. of the objective being used that limits the resolution.

The best resolution from microscopes is summarized below:

	Best Theoretical Resolution	Best Practical Resolution
Light microscopy	0.23 μm	0.29 μm
Electron microscopy	0.2 nm	10 nm

The refractive indices of media used in microscopy are listed below:

Medium	Refractive Index (n)
Air	1.0002
Water	1.3330
Glycerin	1.4700
Immersion oil	1.515

Ocular Micrometers

An ocular micrometer is a ruler in one of the eyepieces of a microscope. It is actually a round piece of glass, called a reticule, into which lines have been etched. If you want to know the length of something in your microscope field, you can use the ocular micrometer. Your cell, for example, may be the same length as 13 divisions of the ocular micrometer; however, if you go up to a higher objective, the cell might be equivalent to 33 divisions of the ocular micrometer. Clearly, something is amiss. The ocular micrometer *looks* the same size in both views, but you know that it is the specimen that *is* the same size. What's up?

What's up is that the image of the specimen has been magnified by the objective, but the image of the micrometer has not. So, when you change objectives, you change the length *represented* by one division of the ocular micrometer. In fact, while the lines in the micrometer are equally spaced, the spacing is arbitrary. To use an ocular micrometer as a meaningful ruler, you must therefore determine what length is represented by an ocular division while using each different objective. In other words, you have to calibrate the ocular divisions, and you have to do it, independently, for *every one* of your objectives. Yes, that's right: If you have five objectives, you have to calibrate the ocular micrometer five times. If you have three microscopes, each of which has four objectives, you will be doing 12 calibrations. If you move the ocular micrometer to a new microscope, you must recalibrate it for each of the objectives on that new microscope.

To determine the length represented by one division of an ocular micrometer for each of the objectives that you have, you use a ruler called a stage micrometer. A stage micrometer is a glass slide with a tiny ruler etched into the middle; the ruler is sometimes surrounded by a black circle to help you find it. The lines of this ruler are equally spaced, and the spacing is decidedly *not* arbitrary; it is very specific, and the exact length will be indicated somewhere on the slide. Most of the stage micrometer ruler will be divided into long lengths, for example, 100 μm between neighboring lines. At one end of the ruler will be a section that is divided into shorter lengths, for example, 10 μm between neighboring lines. The actual lengths between the markings will be etched onto the side of the slide. Usually, every fifth line of a stage micrometer is taller, and every tenth line is taller still. The different lengths of the lines help you to count. You use the stage micrometer as a ruler to measure the distance between lines of the ocular micrometer. This sounds very straightforward, but it is frequently very difficult to do. The following step-by-step instructions may be very helpful.

METHOD **Calibrating an Ocular Micrometer**

1. Adjust the microscope so that the ocular micrometer is in sharp focus.

2. With your smallest objective in place (often 4x or 10x), put the stage micrometer on the stage so that the region with the smaller divisions is to the right, and focus. You must have both the stage and ocular micrometers in focus. To distinguish which micrometer is which, you can rotate the eyepiece; the ocular micrometer will rotate. Line up

the two micrometers so that they are parallel and partially overlapping as in the following figure.

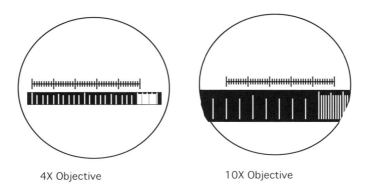

4X Objective   10X Objective

3. Measure the distance between ocular divisions using the stage micrometer as a ruler. Don't measure one individual length, however; measure the distance separating two distant lines. You will divide by the number of ocular divisions separating those distant lines. This is a more precise way to measure, because there is always error associated with measuring, but when you divide, you divide the error as well, and the error associated with the quotient is greatly reduced. Align the left end of the ocular micrometer with a mark on the left end of the stage micrometer. Make sure the distant line will be in the region of the small divisions on the stage micrometer. Count, and write down: (1) the number of ocular divisions, (2) the number of wider stage divisions, and (3) the number of narrower stage divisions.

4. Multiply the number of wide-stage micrometer divisions by the distance that separates them (etched onto the slide; usually 0.1 mm) and multiply the number of narrow-stage micrometer divisions by the distance separating them (etched into the slide; usually 0.01 mm). Add the two products. That sum is the actual distance represented by the two distant lines on the ocular micrometer.

5. Divide the actual distance by the number of ocular divisions separating the distant lines. This tells you the distance represented by a *single* ocular division with your smallest objective in place (see the figure below).

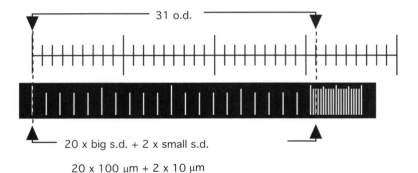

20 x big s.d. + 2 x small s.d.

20 x 100 μm + 2 x 10 μm

31 o.d. = 2020 μm

$$1 \text{ o.d.} = \frac{2020 \text{ μm}}{31 \text{ o.d.}} = 65 \text{ μm}$$

s.d. indicates divisions on the stage micrometer.
o.d. indicates divisions on the ocular micrometer.

6. Repeat this process for the other objectives.

7. Once you have calibrated the ocular micrometer for all the objectives in a particular microscope, it is helpful to create a chart and tape it someplace obvious, such as the base of the microscope. That way, the micrometer does not have to be calibrated again (until you get a new objective or move the micrometer to a new microscope). A sample chart is shown below:

Ocular Micrometer Conversion Factors		
Using 4x objective, 1 o.d.	=	_____ μm
Using 10x objective, 1 o.d.	=	_____ μm
Using 20x objective, 1 o.d.	=	_____ μm
Using 40x objective, 1 o.d.	=	_____ μm
Using 63x objective, 1 o.d.	=	_____ μm
Using 100x objective, 1 o.d.	=	_____ μm

When calibrating an ocular micrometer always double check your numbers by making sure that they follow these patterns:

1. The higher the magnification, the smaller the distance represented by an ocular division.

2. Lengths represented by 1 o.d. in different objectives should have the following relationship to one another: The ratio of the length represented by 1 o.d. seen through the ax objective to the length represented by 1 o.d. seen through the bx objective should be $b \div a$:

$$\frac{1 \text{ o.d. through } a\text{x objective}}{1 \text{ o.d. through } b\text{x objective}} = \frac{b}{a}$$

For example, if 1 o.d. represents 25 μm when viewed through the 10x objective, 1 o.d. viewed through the 4x objective is given by $\frac{25 \text{ μm}}{x} = \frac{4}{10}$, and $x = 62.5$ μm. Likewise, if 1 o.d. represents 25 μm when viewed through the 10x objective, 1 o.d. viewed through the 20x objective is given by $\frac{25 \text{ μm}}{x} = \frac{20}{10}$, and $x = 12.5$ μm.

When you publish or otherwise present a micrograph, it must have a scale bar on it. To make a scale bar, you must have a picture of the stage micrometer taken at the same magnification as your image. So, while you have your stage micrometer handy, you might want to take a picture of the stage micrometer with each of your objectives. Then, whenever you have a micrograph that needs a scale bar, you already have the picture you'll need to make it. Don't forget to make a note of which objective you used for which image.

Once your ocular micrometer is calibrated, whenever you want to measure a length using the microscope, just line up the ocular micrometer with the item to be measured, count the number of ocular divisions, then multiply that number times the actual length for the objective you are using, as listed in your chart. Voila! You know the length.

METHOD Measuring Lengths Using an Ocular Micrometer

1. Make sure the ocular micrometer is in focus.

2. Move the stage so that the specimen you wish to measure is in focus and overlaps the ocular micrometer. The eyepiece can be rotated to align the micrometer with the specimen.

3. Count the number of ocular divisions corresponding to the object you are measuring.

4. Multiply that number by the conversion factor listed in your chart to get the actual length of the object.

Ocular Micrometer Uncertainty

As long as your calibration is good, it is the precision with which you can measure that is going to be the important component of your uncertainty. A safe estimate is that you can distinguish between whole numbers of ocular divisions, although if you have very good eyes, you may feel confident that you can judge within ± half a division. This means that your uncertainty is ±1 o.d. Since the length represented by 1 o.d. changes with the objective, so does your uncertainty. Using the numbers from the above hypothetical calibration, if you are using the 10x objective, your measures will be ±25 μm.

COUNTING CELLS (NUMBER)

Different methods and devices are used to count cells. They vary by sophistication and cost. Hemocytometers are relatively inexpensive, and if you just need a ballpark cell count, they are fine. For example, they are used to get a rough estimate of the percent viability of cells in culture. Hemocytometers, however, are somewhat labor-intensive, depending on

how many counts you need to make. Coulter counters and flow cytometers are expensive, but they are automated and can count large numbers of cells; however, the preparation of the cells in advance of using a flow cytometer is more involved. Flow cytometers can also be used to sort cells.

Hemocytometers

To estimate the concentration of cells in a suspension, you do not need to count every cell; you can count a representative sample and extrapolate. Hemocytometers are made so that a predetermined volume of the suspension can be easily counted. That count is then converted back into a concentration for the original suspension. Hemocytometry is not good for absolute counts of very large numbers of cells. For that, it is more appropriate to use an automated counter, such as a Coulter counter (see below).

Counting the Number of Cells per Milliliter of Suspension with a Hemocytometer

Cells must be evenly suspended and at an approximate concentration of 5×10^5 to 1×10^6 cells per milliliter (50–100 cells per 0.1-µl square). The hemocytometer chamber must be filled exactly, and the coverslip must rest on its supports. The coverslip that comes with the hemocytometer is not your average coverslip and, if it gets broken, can only be replaced by another hemocytometer coverslip. A regular coverslip will not be the correct weight to overcome the surface tension of the fluid drop, and the "chambers" will then be the wrong volume.

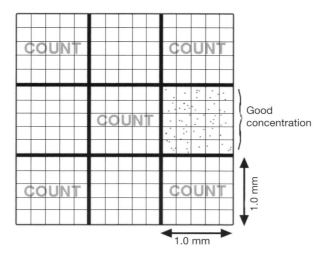

Each of the two chambers of a hemocytometer is divided into nine big squares (see figure above), each of which is 1 mm x 1 mm x 0.1 mm when the coverslip is on. Thus, when loaded properly, each of the nine squares demarcates 0.1 µl of fluid. (Each of the 18 squares is further subdivided, but you can ignore that for now.) To determine the number of cells per microliter, count the cells in 10 of the 18 large squares (the four corner squares and the center square on each side of the hemocytometer), and then calculate the cells per milliliter in the measured suspension. If the original suspension had to be diluted (to give an approximate concentration of 50–100 cells per square), then the number of cells per milliliter of the diluted suspension, times the dilution factor, is the number of cells per milliliter of the original suspension.

METHOD How to Use a Hemocytometer
(See also Plug and Chug in Chapter 8.)

1. Put the special coverslip on the dry hemocytometer.

2. Make sure the cells are evenly suspended.

3. Use a pipetter to pick up a small amount of the evenly suspended cell culture (remember that you are filling chambers that hold less than 1 µl each).

4. Touch a drop of sample to the triangular depression on the side of one chamber, allowing capillary action to fill the chamber. Repeat this process for the second chamber. If there is any fluid remaining in the troughs between the chambers, start over.

5. Put the hemocytometer on the stage of your microscope. Using your 4x objective, confirm that there are about 50 to 100 cells per square. The fewer cells there are, the easier it will be to count. If there are too many cells, make a dilution of the original suspension and start at step 1 using the diluted sample. Keep track of the dilution factor (if you dilute 1:4, the dilution factor is 4).

6. Count the cells in ten of the squares of your hemocytometer (typically the four corner squares and the center square of each chamber). To count cells that touch the lines marking the boundaries of the hemocytometer squares, designate two edges per square as "if touching, count" and the other two as "if touching, do not count." Always choose the same two edges so that you never get mixed up.

7. Sum the ten counts from your hemocytometer to give you the number of cells per microliter. An alternative is to find the mean and the 95% confidence intervals for the ten counts (see Chapter 7). This gives you an estimate of the number of cells per 0.1 µl, plus a measure of your uncertainty.

8. Multiply the sum in step 7 by 10^3 to estimate the number of cells per milliliter. An alternative is to multiply the mean plus or minus the confidence limits by 10^4 to estimate the number of cells per milliliter in the measured suspension.

9. If you are counting a dilution of your original cell culture, multiply your estimate by the dilution factor (if you diluted 1:4, multiply by 4). This gives the number of cells per milliliter in the original suspension.

The relevant equation for determining cells per milliliter:

$$\left(\sum_{i=1}^{10} \text{Cells in square}_i \right) \times 10^3 \times D = \text{cells/ml}$$

Where D = dilution factor; a dilution of $1:D = 1$ part concentrate + $(D-1)$ parts diluent. An alternative equation is:

(Average cells per square ± 95% confidence intervals) × 10^4 × D = cells/ml

☑ Example

You have a suspension of cells in 2 ml of growth medium, and you want to know how many cells you have per milliliter. Consider the following scenario:

1. Following the directions given above, you load both chambers of a hemocytometer.

2. Maybe the first time, you find that there is fluid in the troughs; so you start over.

3. When you put the now perfectly loaded hemocytometer on the microscope stage and look, you observe that there are about 400 cells per square, and this is far too many. So, you wash off and dry the hemocytometer, again, and go back to your suspension of cells.

4. You dilute a small aliquot of your evenly suspended cell solution 1:5 (e.g., you may dilute 10 µl of cell suspension in 40 µl of medium), and you load the hemocytometer with this diluted suspension. Nothing overflows, and there are now about 80 cells per square.

5. You start to count. You get the following cell counts for the ten hemocytometer squares: 70, 83, 72, 72, 90, 79, 75, 81, 67, and 68. The sum of these ten counts is: 757. You estimate, therefore, that there are 757 cells/µl in the diluted suspension. Thus, it follows that there are 757×10^3 cells/ml in the diluted suspension.

> **An alternative is:**
>
> You start to count. You get the following cell counts for the ten small hemocytometer squares: 70, 83, 72, 72, 90, 79, 75, 81, 67, and 68. The mean of those counts is 75.7 and the 95% confidence interval is ±5.3. (If you are statistically inclined, you may want to know that this number was determined by comparison to a *t* distribution, with degrees of freedom (DF) = 9. If you are not yet statistically inclined, Chapter 7 explains how to get that number.) So, the estimate for the number of cells per 0.1 µl is 70.4 to 81.0. Cells only come in whole numbers (and since the original counts have only two significant digits, the estimate should only have two significant digits); therefore, the estimate is 70–81 cells per 0.1 µl. That means that there are 70×10^4 to 81×10^4 cells/ml in the measured suspension.

6. If there are 757×10^3 cells/ml in the suspension that was diluted 1:5, the original suspension is 5x as concentrated, i.e., 757×10^3 cells/ml $\times 5 = 3785 \times 10^3$ cells/ml. The original counts had only two significant figures, so this should be reported as 3.8×10^6 cells/ml. A single summary equation that represents the above example looks like this:

$$(70 + 83 + 72 + 72 + 90 + 79 + 75 + 81 + 67 + 68) \times 10^3 \times 5 = 3.8 \times 10^6 \text{ cells/ml}$$

> **An alternative is:**
>
> If there are between 70×10^4 and 81×10^4 cells/ml in the suspension that was diluted 1:5, the original suspension is 5x as concentrated, i.e., 3.5×10^6 to 4.1×10^6 cells/ml in the original suspension.

Determining Percent Viability Using a Hemocytometer

To determine the percentage of viable cells in a sample, you first treat the cells with a dye that distinguishes live cells from dead cells. For example, trypan blue is excluded by live cells but is taken up by dead cells. In a suspension stained with trypan blue, dead cells therefore turn blue and are easily distinguishable from viable cells, which are unstained. Once the cells are differentially stained, you count as described above: (1) the total number of cells in each of 10 of the 18 large squares of the hemocytometer, and (2) the number of live (unstained) cells in those same 10 squares. It may be easier to keep track of the dead blue cells. The difference between the total and the number that are blue then gives you the number of viable cells in the 10 squares. The percent viability is the number of viable cells in the 10 squares divided by the total number of cells in the 10 squares, times 100.

METHOD **Determining Percent Viability**
(See also Plug and Chug in Chapter 8.)

1. Follow steps 1–6, as described above in "How to Use a Hemocytometer," using a cell suspension treated with a dye such as trypan blue. Keep track of the number of unstained cells, as well as the total number of cells.

   > **An alternative is:**
   >
   > Follow steps 1–6, as described above in "How to Use a Hemocytometer," using a cell suspension treated with a dye such as trypan blue. Keep track of the number of living (i.e., unstained) cells, as well as the total number of cells. For this count, however, you only need the means of the two counts; you do not need the confidence intervals (yet).

2. Divide the number of unstained cells by the total number of cells.

   > **An alternative is:**
   >
   > Divide the mean number of unstained cells by the mean total number of cells.

3. Multiply the quotient by 100. That is the percent viability.

   > **An alternative is:**
   >
   > Multiply the quotient by 100. That is the percent viability. For example, if the mean number of living cells is 62.5 and the mean total number of cells is 76.1, the ratio is 62.5 ÷ 76.1 = 0.82, which is 82%. The uncertainty associated with this number turns out to be somewhat complicated to calculate. This is because the number 0.82 is a proportion, and the data are what is called binomial (which means that there are only two possible results, in this case, living or dead). The 95% confidence limits for this estimate turn out to be 72–91% (which means there is a 95% probability that the actual percent viability is between 72% and 91%). The explanation of how to calculate those numbers is given in Chapter 7.

Coulter Counters

A Coulter counter is a machine used to count cells. The anatomy of a Coulter counter is, essentially, an inner chamber (called the aperture tube) inside an outer chamber (picture a test tube inside a beaker). Both chambers are filled with electrolyte solution (a solution that conducts electricity). The outer chamber has a positively charged electrode suspended in it; the inner chamber has a negatively charged electrode suspended in it. There is a small aperture in the wall of the inner chamber. Current can flow through the aperture, and there is a voltage across the aperture. That voltage is proportional to the volume of conducting electrolyte solution near the aperture (a region called the sensing zone).

A Coulter counter can count cells because cells suspended in the medium in the outer chamber are attracted to the negatively charged inner chamber. To enter the inner chamber, the cells must pass through the aperture. As a cell moves through the aperture it causes a blip (a brief change) in the voltage that occurs when the nonconductive cell briefly displaces the conductive medium. The blips are counted, and the number of blips equals the number of nonconducting particles (hopefully cells) that move through the aperture. Nonconducting dust or debris can also cause a blip in a Coulter counter, so it is very important to keep Coulter counters clean and dust-free.

Fortuitously, the size of the voltage blip is proportional to the volume of the nonconducting particle moving through the aperture, so Coulter counters can also measure particle volume.

Flow Cytometers

Flow cytometers can count and sort cells, but the cells must first be labeled with something fluorescent (antibodies, dyes, or fluorescent proteins). Once the cells are fluorescent, a thin stream of cells suspended in medium is injected down the middle of a fluid-filled tube in the flow cytometer. The cells are then "focused" into single file by the hydrodynamics of the tube, and they pass one by one through a beam of light. The wavelength of the light is set to be the excitation wavelength of the fluorophore used to label the cell (actually, you pick your fluorophores to match the wavelengths available in your machine). At the same position in the tube as the beam of light, there is a light collector that will pick up the wavelength emitted by excitation of the fluorescent label (the emission wavelength). This causes a blip, and the number of blips equals the number of labeled cells.

Flow cytometers can also sort cells that have been labeled with different fluorophores. If the light collector detects the emission wavelength of one fluorophore, the cell is deflected into one container; if the collector detects the other emission wavelength of the second fluorophore, the cell is deflected into a different container.

MEASURING CONCENTRATIONS
USING SPECTROPHOTOMETRY

(See also Plug and Chug in Chapter 8.)

A spectrophotometer is a machine that can shine light of a certain wavelength through a cuvette and measure the amount of light that makes it through to the other side of the cuvette. By knowing how much light went in (the incident light) and how much goes all the way through (the transmitted light), the machine can calculate how much light was absorbed by whatever was in the cuvette. The amount of light absorbed is proportional to the concentration of light-absorbing material in the cuvette. So, if the material whose concentration you wish to know absorbs light of a particular wavelength, you can use a spectrophotometer to measure its concentration.

Put another way: The idea behind calculating concentration from absorbance readings (which are given by a spectrophotomer) is that if you shine a light through a sample of molecules that can absorb light, the number of photons absorbed will be proportional to the concentration of absorbing molecules. Absorption by the sample reduces the number of photons moving in a straight line, thus reducing the intensity of the light emerging undeflected on the other side of the cuvette. The light that gets all the way through the sample without being absorbed (and reemitted in another direction) is called the transmitted light.

The intensity of transmitted light (called I) is inversely proportional to the number of absorbing molecules in the sample (c, measured as concentration in M). Also relevant are the amount of light going straight into the sample (called the intensity of incident light, I_0); the distance the light has to travel through the sample (called d, the optical path length; $d = 1$ cm); and the ability of the material to absorb light (called the molar extinction coefficient or molar absorption coefficient, ε). ε equals the absorbance of a 1 M solution with $d = 1$ cm; thus, its units are M^{-1} cm^{-1}. The relationship among those variables is $I = I_0 10^{-\varepsilon dc}$.

The Beer-Lambert Law

What you actually measure is the absorbance, A. The Beer-Lambert law defines A:

$$A = -\log \tfrac{I}{I_0} = -\log 10^{-\varepsilon dc} = \varepsilon dc$$

A is dimensionless. When $d = 1$ cm, which it practically always does, then A is called the optical density or OD. This relationship, $A = \varepsilon dc$ (i.e., the Beer-Lambert law), should make you happy, because it says that the relationship between the absorbance and the concentration of absorbing molecules is linear. (You can think of the generic equation for a line: $y = mx + b$; in this case, y is A, m is εd, x is c, and b is 0.) Linear relationships are easy to deal with; so, be happy.

Practical Considerations

Consider the following:

1. Light-absorbing molecules absorb different wavelengths to different extents. For example, protein absorbs light of many wavelengths, but it absorbs 280-nm light *best*. To determine what wavelength your molecule absorbs best, measure A at a wide range of wavelengths, and then graph A as a function of wavelength (some spectrophotometers will do this automatically). Wherever the peak is, that wavelength is what the molecule absorbs best. Once you know what wavelength is maximally absorbed, measure and report the A (or OD) for that wavelength. The A at a particular wavelength is written A_λ (or OD_λ).

2. $A_\lambda = \varepsilon dc$, the Beer-Lambert law, relates the absorbance (i.e., the optical density) to the concentration of molecules present in molar concentrations. If measured concentrations are, in fact, in the molar range, the extinction coefficient ε has units of M^{-1} cm^{-1}. If, however, you are dealing with a solute that is likely to be present in millimolar concentrations, you use $A_\lambda = Edc$, where E is the millimolar extinction coefficient, which has units of mM^{-1} cm^{-1}.

3. The Beer-Lambert law is useful if you are trying to determine the extinction coefficient (molar or millimolar) of your molecule. To use it for that purpose, you measure a few solutions with known concentrations and then rearrange the law to:

$$\varepsilon = \frac{A_\lambda}{dc}$$

You read A_λ off the spectrophotometer, d almost always equals 1 cm, and you know c a priori (because you used solutions of known concentrations). Average your values, and voila, you know ε (or E).

4. If you know the wavelength of maximum absorption (λ_{max}) and the extinction coefficient (ε or E) for your molecule, determining the concentration of those molecules in solution is a simple matter, as long as the solution is definitely not contaminated with some other molecule that might absorb λ_{max}. All you need to do is allow the spectrophotometer to warm up for 15 minutes, zero it, and then measure $A_{\lambda max}$. Then plug the numbers into one of the following versions of the Beer-Lambert law, and solve for concentration:

$$c = \frac{OD_\lambda}{d\varepsilon}$$

c = Concentration [M]
OD_λ = A_λ = Optical density at λ_{max} for the molecule. Read this number off the spectrophotometer.
d = Width of the cuvette [cm]; almost always 1 cm.
ε = Molar extinction coefficient [M^{-1} cm^{-1}]. Look this number up.

$$c = \frac{OD_\lambda}{dE}$$

c = Concentration [mM]
OD_λ = A_λ = Optical density at λ_{max} for the molecule. Read this number off the spectrophotometer.
d = Width of the cuvette [cm]; almost always 1 cm.
E = Millimolar extinction coefficient [mM^{-1} cm^{-1}]. Look this number up.

5. If you are trying to determine the concentration of something, but don't know the extinction coefficient, you can still get the concentration. In this case, the Beer-Lambert law is something you know and believe; but you don't need to use it because you can make a standard curve that directly relates OD_λ to c and you can use that curve to tell you the concentration of your unknowns. As with any calibration, pick standards (your samples of known concentration) that are as similar as possible to your molecule, and pick the concentrations such that they bracket the unknown concentrations you are likely to be measuring. Allow the spectrophotometer to warm up for 15 minutes and zero it. Then measure $A_{\lambda max}$ of one standard, then zero, then measure $A_{\lambda max}$ of the next standard, then zero, then measure $A_{\lambda max}$ of the next standard, etc. Graph the absorbance as a function of concentration; this is your standard curve. Now, after zeroing the spectrophotometer, measure the absorbance of your unknown, and read its concentration off your standard curve.

SEPARATING SAMPLE COMPONENTS
USING CENTRIFUGES

The centrifuge is a frequently used piece of equipment in a laboratory. It is used to pellet cells out of suspension, for subcellular fractionation, and as a part of many routine protocols such as those for protein or DNA isolation.

The idea of centrifugation is to spin a sample very fast, so that it experiences the effects of inertia and centripetal forces (see explanations below). Components of the sample that differ in certain physical properties, such as mass, will move differently in response to the forces and thus will end up in distinct places within their container; i.e., they will be physically separated. For a centrifuge to work properly, weight in the centrifuge rotor *must* be distributed evenly. Tubes and volumes of liquid, for example, must be balanced. This is an important safety issue, because the forces generated by a centrifuge can be dangerous to you, not to mention your expensive centrifuge.

As mentioned above, two forces are acting on the components of a sample inside a tube in a spinning centrifuge. Centrifugal force, however, is not one of them. Centrifugal force doesn't actually exist: There is no force that pushes outward on the contents of a spinning tube. There are two real forces acting on objects in a centrifuge. The first is inertia. Each component in a centrifuge is essentially "trying" to travel in a straight line due to its inertia. The second force is the inward, or centripetal, force on the sample that is attributable to the tube being attached to the center of the machine (centripetal means center-seeking). This inward force bends the path of the sample into a circle.

Centrifugation depends on different components of a sample having different masses, and therefore different amounts of inertia. Items with more mass and hence more inertia will be able to take a path that is straighter than the path of the items with less inertia. Within a revolving tube, the straighter path is found further away from the axis of the revolution, because the greater the radius of the circle traced, the straighter the path of the particle tracing it. So, the denser components of a sample move further into the tube, whereas the less dense items end up closer to the top.

There are many different types of centrifuges, such as fixed-angle and swinging-bucket, and many methods of centrifugation, such as differential centrifugation and density gradient centrifugation. Each type exploits various physical phenomena to separate samples based on different physical properties.

How to Use the Centrifuge You Have to Get the Separation You Need

If you wish to use a protocol that includes a centrifugation step, you need to figure out how to get the right acceleration. Acceleration in a protocol may be referred to in terms of g-force or gs, and this is written as some number followed by a g, for example "centrifuge for 15 minutes at 1500g." This nomenclature relates the acceleration within the centrifuge to gravitational acceleration, which is a constant symbolized by the letter g; you can think of g as a unit for the property acceleration. Confusingly, however, g-force is not a force, and the accelerations indicated in this way are not multiples of the gravitational constant. Nevertheless, it is how the accelerations in a centrifuge are reported, so that is what you are attempting to accomplish with your centrifuge. Some centrifuges will set themselves to deliver the appropriate number of gs if you input the acceleration you want. Others, however, require you to determine how to accomplish the necessary gs. You may therefore have to calibrate your centrifuge.

On many centrifuges, you cannot directly change the gs, also known as the relative centrifugal field, RCF. What you can change on a centrifuge is the rotations per minute (rpm). The relationship between the rotations per minute and the relative centrifugal field (i.e., the number of gs) is given by the following equation:

$$RCF = 1.1 \times 10^{-6} \times (rpm)^2 \times radius$$

RCF = Relative centrifugal field (dimensions LT^{-2}; reported using g as a unit)

rpm = Rotations per minute [min^{-1}]

Radius = Distance from the center of the rotor to a relevant position in the tube [mm]

In this equation, the radius of interest is the distance from the center of the axis of rotation to the relevant position in the spinning tube—often either (1) the point in the sample that is furthest away from the axis, which will tell you the maximum acceleration, experienced at the tip of the tube, or (2) the point halfway between the furthest and the nearest points within the sample, which will tell you the mean acceleration experienced by the whole tube. The information may be provided by the manufacturer of the centrifuge, or you may have to measure it yourself. Once you know the radii of the standard rotors for your centrifuge, it is very useful to write the numbers down and store them in an obvious place so that no one else has to go through the measuring.

How the numbers on the speed setting dial on a centrifuge translate to rpm is described in the literature that comes with the centrifuge. In many labs, this information is photocopied and taped to the wall near the centrifuge. Even better, the RCF for each number on the dial is often calculated and then taped to the wall. You can also have your centrifuge calibrated professionally.

To match the rpm to a particular RCF (number of gs), you can either look at a chart that someone has kindly posted near your centrifuge, or you can divide the value of the RCF you are trying to accomplish by the radius (in mm), divide that number by 1.110^{-6}, take the square root of that number, and choose the setting that corresponds to that rpm. In other words, solve this equation:

$$\text{rpm} = \sqrt{\frac{\text{RCF}}{\text{radius} \times 1.1 \times 10^{-6}}}$$

Then, run the centrifuge at the setting that delivers that rpm. Another way to figure out what setting to use (i.e., what rpm to use) is to use a chart called a nomogram; this is a graphical representation of the relationship between rpm, RCF, and rotor diameter.

To use the nomogram to determine the rpm, draw a straight line from the radius of the rotor, through the RCF you want, to the rpm line. This will tell you the rpm needed to achieve the gs you want.

MEASURING RADIOACTIVITY—NUCLEAR DECAY

A very important piece of equipment in a laboratory that uses radioactive isotopes is the Geiger counter. Although most people know from experience what to listen for when using a Geiger counter, monitoring radiation is a serious task; so the more you know the better. This section is not meant, in any way, to replace appropriate radiation safety training provided by your institution.

Geiger-Muller Counter

A Geiger-Muller (GM) counter, named for Hans Geiger and W. Mueller, who invented this device in the 1920s, is used to detect radioactivity. A GM counter works on the principle that radiation passing through a gas will ionize some of the atoms in the gas. The ions will subsequently affect the current passing through a wire running through the gas, and this change in current can be converted to a measure of radioactivity. GM counters can pick up small amounts of radiation, but only if that radia-

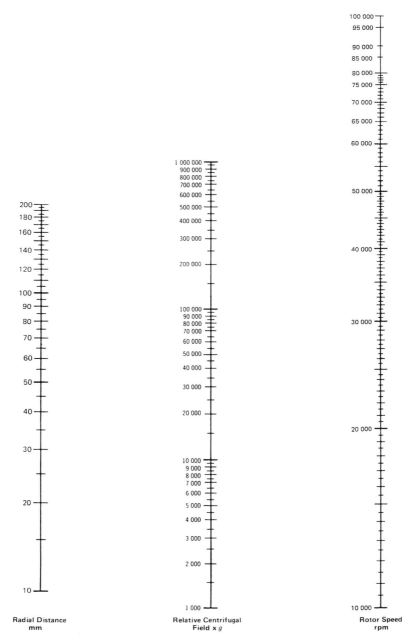

Nomogram for computing relative centrifugal forces. (Reprinted, with permission, from Corning Life Sciences, Acton, Massachusetts.)

tion is of high enough energy to penetrate into the gas-filled chamber of the counter. For example, most alpha emitters, such as ^{222}Rn, are not strong enough to trigger a reading on a GM unless they are very concentrated.

Different GM counters report results in different units. Some give the measure of radiation in Becquerels (disintegrations per second), others in Sieverts per hour (a measure of dosage rate), and still others in Roentgens per hour (a measure of intensity rate). Typically, measures of radioactivity from a GM counter are accompanied by a sound. The frequency of the blip of a Geiger counter is proportional to the amount of radiation being measured. This feature of a Geiger counter makes it an invaluable safety tool because it enables you to focus your eyes on your work while you monitor with your ears.

Scintillation Detector

Another type of very sensitive radiation detector is the scintillation detector. Scintillation detectors exploit materials (commonly sodium-iodide salt) that emit photons when hit by radiation. The photon signal is amplified by a photomultiplier tube, and the amplified signal is detected and converted into a measure of radioactivity. Scintillation counters are more sensitive than Geiger counters because they can be set to detect many types of radiation, including emitters that are too weak to be detected by a GM counter. They also give a more precise measure of radioactivity than a GM counter, which makes scintillation counters the machine of choice for measuring the incorporation of radioactive isotopes into probe molecules.

Another important use of the scintillation counter is to detect any radioactivity in the swipes collected during a weekly check for radioactive contamination. Again, be sure to follow the radiation safety procedures appropriate for your laboratory.

RESOURCES

Koehler Illumination

http://www.zeiss.de

Geiger-Muller Counter

http://www.ccohs.ca/oshanswers/phys_agents/ionizing.html

Making Solutions

Good solution making is an important skill for scientists because having properly made solutions is critically important to the success of any experiment. Solution making is also one of the least glamorous of lab tasks, and it is frequently passed to someone else. Unless you are absolutely sure that your solution maker is skilled with math, not making your own solutions can be very risky. An incorrectly made solution can ruin an experiment. Furthermore, when something does go wrong with an experiment, determining that the solution is the source of the problem can be arduous.

The calculations involved in making solutions have confused and frustrated legions of science students and have led to vast piles of calculation-covered paper scraps that radiate a desperate hope of stumbling across numerical revelation. Calculations lurk in almost every step of solution making: creating a recipe from a list of concentrations; calculating moles and molarity from masses, formula weights, and volumes; measuring accurately and determining the precision of the measurement; calibrating and using a pH meter; making aliquots; and diluting.

This chapter (which, by the way, was the first written and remains the most appreciated) addresses the mathematical hurdles of solution making in two ways. First, there are short explanations and reminders of what you are doing when you are making a solution and why you are doing it. Second, there are "How To Quickly" shaded boxes. If you don't have time, at the moment of truth, to understand the rationale for the math you are doing (you will come back later and learn, of course), you can find out "How To Quickly" do the necessary calculations to get on with your work.

TALKING ABOUT SOLUTIONS

Acid: The classical (Arrhenius') definition of an acid is a molecule that dissociates in water to yield H_3O^+. The Brønsted-Lowry definition of an acid is a molecule that can donate (release) a proton in the form of H^+.

Base: The classical (Arrhenius') definition of a base is a molecule that dissociates in water to yield OH⁻. The Brønsted-Lowry definition of a base is a molecule that can accept a proton in the form of a hydrogen atom (H⁺).

Acid Dissociation Constant, K_a*:* The equilibrium constant of an acid.

$$K_a = \frac{[H]^+[OH^-]}{[H_2O]}$$

If $K_a > 1$, it means the acid is strong, i.e., likely to readily donate a proton. If $K_a < 1$, it means the acid is weak, i.e., less likely to donate a proton.

Buffer: A buffer is a weak acid that is added to a solution to prevent changes in the pH when a small amount of a strong acid or base is added. Buffers work by accommodating the added acid or base with changes in the relative concentrations of the weak acid and its conjugate base (i.e., the concentration of buffer molecules that can donate protons and the concentration that can accept protons).

cc: cc stands for cubic centimeter. One cc equals one milliliter.

Diluent: The diluent is the medium or solvent added to a concentrated solution in order to dilute it.

Diluting 1:X: Recipes for solutions sometimes contain directions for diluting a stock solution according to a certain ratio. To read those directions, you need to know the following. A 1:X dilution means your concentrated solution should be diluted to 1/Xth its current concentration. Add 1 volume of concentrate to ($X - 1$) volumes of diluent to create a total volume equal to X. Looking at it yet another way, in the final solution what you diluted will be 1/Xth of the total volume. Therefore:

1:100   means 1 part concentrate, 99 parts diluent
1:14   means 1 part concentrate, 13 parts diluent
1:2   means 1 part concentrate, 1 part diluent
1:1   means straight concentrate

Note: This is the original convention; however, it is not always followed. If you see 1:9, chances are the author means 1:10 and the recipe should be checked. This is particularly true if you see dilute 1:1 in a recipe; chances are the author means 1:2. You have to use your judgment. This confusion is one reason why it is always better to report starting and final concentrations explicitly.

λ: λ is an abbreviation for microliter (μl) and the symbol for wavelength. The context around the use of λ indicates which meaning is applicable to the situation.

Medium: A solution. The plural is media. In biology, the word "medium" is typically reserved for solutions made specifically to feed cells.

Meniscus: The surface of the liquid in a thin cylindrical container is concave, and the rounded surface is called the meniscus. Glassware is calibrated so that the correct volume is measured when the bottom of the concave meniscus lines up with the volume marking.

Molarity: The molarity of a solution is the moles of solute per liter of solvent. Molarity is an SI unit and the symbol is M. If there is more than one component in a solution, the molarity of each is given separately. For example, a solution might be 1 M NaCl (58.43 g/mole; so the solution has 58.43 g of NaCl per liter of solution), 0.5 M $CaCl_2$ (219.08 g/mole; so the solution has 109.54 g of $CaCl_2$ per liter of solution).

Molar: Molar is an adjective describing molarity. For example, a bottle is labeled 1 M $MgCl_2$; the solution is "one molar magnesium chloride," which means that the molarity of $MgCl_2$ in the solution is one mole of $MgCl_2$ per liter of solution.

Molality: The molality of a solution is the moles of solute per kilogram of solvent. Molality is not really analogous to molarity: Not only do you substitute kilograms for liters, but molarity is moles per liter of *solution*, whereas molality is moles per kilogram of *solvent*. The symbol for the unit is a lowercase *m*.

Mole: An amount; 6.022×10^{23} of molecules of a substance. One mole of a substance has a mass equal to its molecular weight in grams. For example, HEPES has a molecular weight of 238.3; one mole of HEPES "weighs" (has a mass of) 238.3 grams. Mole is the SI unit for amount of substance. It is abbreviated mol.

Percent: Percent is a dimensionless parameter meaning per one hundred (as in *per centum*); 37% of $A = 37/100 \times A = 0.37A$.

Percent Volume per Volume (v/v) or Weight per Weight (w/w): The amounts of ingredients in a solution are sometimes described as a percentage of the total solution. To read those recipes, you need to know the following. If the units are the same (e.g., w/w or v/v), the percentage is what you expect based on the definition of percent:

100% = 1 g/1 g
  1% = 10 ml/liter

Percent Weight per Volume: This is one of those cases where "weight" really means "mass." Percent mass per volume (abbreviated % w/v) is based on a convention which is the mass/volume of pure water: 1 ml of water has a mass of 1 gram; hence, 1 g/ml = 100%. It works as follows:

100% = 1 g/ml
10% = 100 mg/ml
1% = 10 mg/ml

If you just see % without any units such as w/v, you can assume it means w/v.

pH: pH is used to quantify the acidity of a solution. The "p" indicates that you take the negative log of what follows, which is H, signifying the concentration of protons. pH is therefore a measure of the concentration of protons, or H^+, in a solution. As the number of protons in solution increases, the magnitude of the pH decreases; so the lower the pH of a solution, the more acidic it is.

Q.S.: Q.S. stands for "quantity sufficient" and means "add enough solvent to bring the total volume to..." For example, "Q.S. to 500 ml" means "whatever volume you have now, add the amount necessary to make the total volume 500 ml." To Q.S. a solution properly, it must be measured in an appropriately calibrated vessel (i.e., a graduated cylinder, not a beaker).

Reagent: A compound or solution that will go into your reaction mixture.

Solute: The dissolved, or dispersed (as opposed to the liquid), phase of a solution—the stuff you mixed in.

Solution: A homogeneous mixture (usually liquid) of two or more substances; one or more solutes dissolved in solvent. The solutes can be solid, liquid, or gas; the solvent can be liquid or gas.

Solvent: The dispersing (i.e., liquid) phase of a solution. The substance into which a solute is dissolved—what you mixed stuff into.

NUMBERS FOUND ON CHEMICAL BOTTLES

The first thing you are likely to pick up when you begin to make a solution is a bottle of some chemical. Important numbers are found on the labels of chemical bottles and in chemical company catalog listings. Many of the numbers included on chemical bottles have obvious meanings, and others do not; but each provides information that may be extremely useful. Below is a list of numbers typically reported on chemical containers, and why you want to know them.

Product Number: This is the company catalog number, and it is very useful for correct reordering, particularly when a chemical is available in several forms. (The compound may be available in a hydrated form or at a different concentration or purity.)

Lot Number: It is a good idea to record the lot number of a reagent in your lab notebook, especially when using reagents that may vary significantly from batch to batch, such as antibodies or enzymes. Having the lot number easily accessible can be very useful for troubleshooting your experiments. It may be a reagent, not you, that causes an experiment to go awry. When you call a company to ask if a batch of enzyme is known to have been contaminated or is of low activity, for example, technical support at the company will need to know the lot number.

Storage Temperature: This tells you the temperature to store a reagent to avoid things like loss of activity or decay. Pay attention to the storage temperature (and expiration dates) of chemicals, because chemicals really can go bad.

–70°C	means keep it in the ultracold freezer (usually –70°C to –80°C)
–20°C	means keep it in the deep freeze
–20°C–0°C	means keep it in the freezer compartment of the refrigerator
0°C	means keep it in the freezer compartment of the refrigerator
2°C–8°C	means keep it in the refrigerator
RT	means keep it at room temperature

Also watch labels for notes indicating whether a chemical should be stored desiccated or in the dark.

Formula Weight (FW): This tells you the mass in grams per mole of the chemical as listed in the formula on the label; in this context, FW is being used interchangeably with molecular weight. If it is a hydrated compound, the FW includes the mass of the water. For more on this topic, including FW versus MW, see Chapter 2.

Purity: Sometimes trace amounts of by-products of the method used to isolate a chemical compound cannot be entirely eliminated. Purity is a measure of, well, the purity of a chemical, i.e., it tells you how much is not contamination. Purity is typically represented as a percent: The percentage of what you bought that is actually what you want.

Risk and Safety Numbers: These numbers refer to lists found in the catalogs of companies that sell chemicals. They are like mini-Material Safety Data Sheets (MSDS); but the MSDS is the only reliable source for complete safety information. MSDSs are available free online at a number of sites, as listed in Resources at the end of this chapter.

Activity and Unit Definition: Activity and unit in this context apply to enzymes and tell you how much activity to expect from an enzyme and what the units of that activity mean. For example, alcohol dehydrogenase can be purchased with the activity 300–500 units per milligram protein, where 1 unit will convert 1.0 µmole of ethanol to acetaldehyde per minute at pH 8.8 at 25°C. In this case, the activity is a rate.

MOLES AND MOLARITY

Moles and molarity are both units (see Chapter 1). Mole is the SI unit for the dimension of an amount of a substance (dimension N), and molarity is the SI unit for concentration, where concentration is the amount of a substance per liter of solution (dimension NL^{-3}). So, the number of molecules of a chemical is counted in moles, and the concentration of molecules in a solution is quantified as the molarity.

The word "mole" is like the word "dozen" in that it is a special name for a particular number. Dozen is a name for the quantity 12; mole is a name for the quantity 6.022×10^{23}. 6.022×10^{23} goes one better than 12 in that it actually has *two* special names. One of the names, at the risk of being repetitive, is "mole" and the other is "Avogadro's number" (in honor of Amedeo Avogadro, who in 1811 was the first to distinguish between atoms and molecules, and who stated that equal volumes of all gases at the same temperature and pressure contain the same number of molecules, now known as Avogadro's principle). When you talk about moles of a chemical, you are talking about the number of smallest indivisible units of that chemical, i.e., the number of molecules of that chemical. A mole of a chemical is 6.022×10^{23} molecules, just like a dozen eggs is 12 "smallest indivisible units of eggness."

Why people talk about dozens of things is anyone's guess, but there is a very good reason to talk about moles of chemicals: One mole of a chemical has a mass equal to the molecular weight. To figure out how many moles of a chemical you have, you measure the mass of the chemical (using a balance) and then divide that mass by the molecular weight of the chemical (which is listed as the FW on the label). The result is the number of moles you have. For example, HEPES has an FW of 238.3; one mole of HEPES weighs (has a mass of) 238.3 grams. If you weigh out 327.4 g of HEPES and want to know how many moles that is, you simply divide the amount you've weighed by the weight of one mole: 327.4 g ÷ 238.3 g/mole = 1.37 moles of HEPES.

The molarity of a solution is the moles of solute per liter of solvent. Molarity is the SI unit of concentration, and saying "molarity" is the same as saying "moles of solute per liter of solution." The symbol for molarity is M. Molar is an adjective describing molarity. For example, a bottle is labeled 1 M $MgCl_2$; the solution is "one molar magnesium chloride,"

"IT MAY VERY WELL BRING ABOUT
IMMORTALITY, BUT IT WILL TAKE
FOREVER TO TEST IT."

which means that there is one mole of $MgCl_2$ per liter of solution (or one millimole of $MgCl_2$ per milliliter of solution, or one micromole per microliter, etc.). Another way to say it is "the molarity of the solution is 1." If there is more than one component in a solution, the molarity of each is given separately.

MAKING SOLUTIONS FROM DRY CHEMICALS

General Warning
Sometimes recipes will say that upon completion, a solution will be of a certain pH. It is important to check that the solution has actually achieved the noted pH. Sometimes things such as the pH of the distilled water used to make solutions can vary from lab to lab; you cannot blindly count on it to be pH 7. Such factors can alter your ultimate pH.

Solution making most typically involves dissolving a dry chemical into water or other specified solvent. The amount of chemical you add to a solvent depends on the final concentration (molarity) you want in the finished solution and the total amount (liters) of solution you have decided to make. The easiest way to measure chemicals, however, is by mass, since mass is what lab balances report (for more about balances, see Chapter 2). So, to make a solution, typically you have to determine the mass of chemical that you need, based on knowing the desired final concentration (usually molarity), the molecular weight of the chemical, and the final volume of solution. Here is how to do that.

When making any solution, you should start by putting about 80% of the total volume of solvent into your mixing vessel and then add the chemicals one by one. Once all the chemicals have been added, transfer your almost finished solution to a graduated cylinder; then Q.S. (see page 96) to the final total volume. You do it that way because the chemical you are adding will contribute some volume to the solution and you don't want to inadvertently overshoot the desired final total volume.

METHOD **Planning a Solution of a Particular Molarity**

One simple way to plan the making of a solution from a dry chemical is as follows:

1. Figure out the mass you would need if making a full liter.

2. Figure out what fraction of a liter you are making.

3. Use that same fraction of the mass of the chemical.

☑ **Example**

This example shows how to figure out a recipe step by step. At the end of this example is the "How to Quickly" that you can use for reference to get on quickly with making your solution. Say you want 100 ml of a 5.00 M stock solution of calcium chloride, $CaCl_2$. The MW of $CaCl_2$ is 111.0 grams per mole.

1. Figure out how many grams of $CaCl_2$ would go in 1 liter by multiplying the number of moles in 1 liter times the molecular weight of the compound:

   5.00 M = 5.00 moles/liter

   $$5.00 \text{ moles } \times \frac{111.0 \text{ g}}{\text{mole}} = 555.0 \text{ g}$$

2. Figure out what fraction of 1 liter you are making by dividing the desired volume by 1 liter:

   $$\frac{100 \text{ ml}}{1000 \text{ ml}} = 0.1$$

3. Multiply the fraction of a liter times the number of grams to make 1 liter of solution of the same molarity (0.1 × 555.5 g).

0.1 × 555.0 g = 55.5 g

To make 100 ml of a 5.00 M solution of $CaCl_2$, dissolve 55.5 g into approximately 80–90 ml of water and then Q.S. to 100 ml. Written as one expression, this looks like:

$$5.00 \text{ moles} \times \frac{110.0 \text{ g}}{\text{mole}} \times \frac{10^{-1} \text{ liter}}{\text{liter}} = 55.5 \text{ g}$$

This expression can be simplified (which is always good). First, notice that the second term is the molecular weight of the chemical; so, substitute MW. Second, notice that the third term can be written as 10^{-1} liter x liter^{-1}. Since multiplication is commutative, the x liter^{-1} can be moved over and combined with the first term; that makes the first term 5.00 moles/liter, which, you may notice, is the desired molarity, 5.00 M. So the above equation simplifies to:

$$5.00 \text{ M} \times \text{MW} \times 10^{-1} \text{ liter} = 55.5 \text{ g}$$

or, the desired molarity (mole/liter) times the molecular weight (gram/mole) times the desired volume (liter) equals the grams of chemical required.

DON'T FORGET: When you are actually making up the solution, put only about 80% of the total volume of solvent into the beaker before adding the chemicals. Once the chemicals have been added and dissolved, only then should you transfer the contents to a graduated cylinder and Q.S. to the final total volume.

HOW TO QUICKLY Calculate the Grams of Chemical Needed for a Solution of Particular Molarity

Plug your numbers into the following equation (calculate individually for each solute):

M × MW × $V = g$

Add chemicals to 80% of volume then Q.S. to the final volume.

    M = Final molarity of compound [mole liter^{-1}]
  MW = Molecular weight of compound (listed on the bottle as FW) [g mole^{-1}]
    V = Desired final volume [liter]
    g = Grams of compound to add to solvent [g]
 Q.S. = "Add sufficient solvent to bring the volume (to the final volume)"

See also Plug and Chug in Chapter 8.

MAKING SOLUTIONS FROM HYDRATED COMPOUNDS

Some chemicals come with water molecules attached. The first thing to remember is that the MW (listed as FW on the bottle) of such a compound includes the mass of the water; whenever you would have used the MW of the unhydrated compound in the equations above, use, instead, the MW of the hydrated compound. If you have a recipe that tells you how many grams to use of the unhydrated compound, figure out target concentration (see "Calculating Concentrations from Recipes" below) and then calculate the grams to use of hydrated compound.

The second thing to remember when using a hydrated compound is that the attached water molecules contribute water to the solution, potentially diluting the final concentrations (if the solvent is water). Therefore, you must account for the contribution of water from the hydrated compound when determining the volume of solvent (water) to add.

☑ Example 1

1. You can buy sodium phosphate as $NaH_2PO_4 \cdot H_2O$ (sodium phosphate monobasic, FW = 137.99).

2. If you want the molarity of the NaH_2PO_4 to be 0.5 M in a liter, weigh out 68.99 g (i.e., 0.5 × 137.99) and add it to your solution.

   Here's the catch: for every 0.5 mole of $NaH_2PO_4 \cdot H_2O$ you add to the solution, you are adding 0.5 mole of sodium phosphate *and* you are adding 0.5 mole of water. One mole of water takes up 18.015 ml. Just by adding 68.99 g of compound, you have added 9.008 ml of solvent; you need 991 ml (100 – 9.008) more to reach 1 liter.

☑ Example 2

1. You can buy sodium phosphate hydrated with 12 waters: $Na_3PO_4 \cdot 12H_2O$ (sodium phosphate tribasic, FW = 380.12).

2. If you want the molarity of Na_3PO_4 to be 0.5 M in a liter, weigh out 190.06 g of $Na_3PO_4 \cdot 12H_2O$ (0.5 × 380.12) and add it to your solution.

   If you are using this compound, for every 0.5 mole of sodium phosphate you add, you are adding 6 moles of water, which has a volume of 108.09 ml. To Q.S. to 1 liter only requires 891.91 ml (1000 – 108.09).

The moral of these stories is that you must know in advance what volume of water will be contributed by the hydrated compound, and you must accommodate that volume. One way to accommodate the volume of water contributed by a hydrated compound is to calculate exactly what volume of water you will be adding when you add the hydrated compound, and then subtract that volume from the volume of water you planned to add. Using the second of the above examples, if you wanted 100 ml of 0.5 M Na_3PO_4, you could calculate that if you use the $Na_3PO_4 \cdot 12H_2O$, you will be adding 10.81 ml of water (1/10th the volume of 6 moles of H_2O calculated above) along with the compound; so you will only need to add 89.19 ml (100 − 10.81) more water. The steps of that calculation are:

1. Multiply the desired molarity times the desired volume. This tells you the moles of compound you will be adding.

2. Multiply the number of moles of the compound times the number of moles of H_2O in the compound. The result is the number of moles of H_2O you will be adding.

3. The number of moles you are adding, times 18.015, tells you the number of milliliters of water you will be adding.

4. Subtract the number of milliliters you are adding from the total volume.

HOW TO QUICKLY Calculate Water Contributed by a Hydrated Compound

Plug your numbers into the following equation:

M × V × #H_2O × 18.015 = ml H_2O contributed by hydrated compound

Calculate individually for each hydrated solute.

$$
\begin{aligned}
\text{M} &= \text{Desired molarity of compound [mole liter}^{-1}] \\
V &= \text{Desired final volume of solution [liter]} \\
\text{\#}H_2O &= \text{Number of } H_2O \text{ molecules in the hydrated compound} \\
18.015 &= \text{Milliliters per mole of } H_2O \text{ [ml/mole]}
\end{aligned}
$$

See also Plug and Chug in Chapter 8.

The calculation to figure out how much water you are adding is useful for making sure your measuring container doesn't overflow. The easier way to deal with hydrated compounds is to follow the directions for making any solution: Start with less than the total volume of solvent, mix in the chemicals, then transfer it all to a graduated cylinder, and Q.S. to the final volume. When working with hydrated compounds, however, you have to be particularly careful how much you start with. It is a good idea to do the above calculation, and then start with 80% of what you will end up adding (rather than 80% of the total final volume).

HOW TO QUICKLY Make a Solution Using a Hydrated Compound

1. Calculate about how much water will be needed after the hydrated compounds are added using How To Quickly calculate water contributed by a hydrated compound: $M \times V \times \#H_2O \times 18.015 = ml\ H_2O$ contributed. Subtract that many milliliters from the desired final volume (V).

2. Put 80% of the resulting volume of H_2O into a mixing vessel.

3. Add your compounds. Mix well.

4. Transfer the solution-so-far to a graduated cylinder.

5. Q.S. to the desired final volume in a graduated cylinder.

MAKING STOCK SOLUTIONS

In many labs, or for particular experiments, certain solutions are used frequently and are therefore made up in large quantities. To minimize the volume actually occupied by these solutions, they are often made at a higher concentration than that which will be used. These concentrated solutions are referred to as stock solutions. Stock solutions save time, in addition to space, because when you need a solution stored as a concentrated stock, you need only to dilute the stock; you don't have to start from scratch. Some commonly encountered stock solutions are: 10x TBE (in labs that run a lot of gels), 10x SSC (in labs that do lots of hybridizations), and 10x PBS (in labs that do just about anything with cells).

Making a stock solution is just like making any solution; you simply scale up the concentration of all of the ingredients. Typical stock solutions are 2x, 5x, and 10x, meaning double the normal concentration, five times the normal concentration, and ten times the normal concentration. The upper limit to the concentration of a stock solution is set by the solubility of those chemicals in that solvent.

It is important to be very careful not to contaminate a stock solution. If a stock solution gets contaminated, you *must* dispose of it *immediately*. If you do not, you *will* mess up experiments for what will feel like an eternity. To avoid contaminating a stock solution, *never* put *anything* into its container—no pipettes, nothing. Always pour some stock out into a second, clean container, and then take what you need from there. This is just being considerate to others and preventing headaches for yourself.

Diluting a Stock Solution to a Particular Concentration

Having concentrated stock solutions available in the lab is a great time saver if everyone who will be using them is comfortable diluting. The following instructions make diluting easy. The section begins with an explanation, and ends with a How To Quickly.

If you know what concentration of a solution you want and need to figure out how much of a concentrated stock solution to dilute, use the following equation:

$$\frac{\text{Concentration you want}}{\text{Concentration you have}} \times \text{Final volume} = \begin{array}{c} \text{Volume of concentrated stock to add} \\ \text{to mixture} \end{array}$$

You may recognize this as a rearrangement of $M_1V_1 = M_2V_2$, where $M =$ molarity and $V =$ volume. You can use this relationship to calculate everything but the solvent volume, and then Q.S. with solvent to the final volume.

☑ **Example**

You want 25 ml of the following solution:

0.5 M $CaCl_2$
1.0 M $MgSO_4$

You have the following stock solutions:

5.0 M $CaCl_2$
2.5 M $MgSO_4$

Step 1. Figure out how much of the $CaCl_2$ stock to add:

$$\frac{0.5 \text{ M}}{5.0 \text{ M}} \times 25 \text{ ml} = 2.5 \text{ ml of } CaCl_2 \text{ stock}$$

Step 2. Figure out how much of the $MgSO_4$ stock to add:

$$\frac{1.0 \text{ M}}{2.5 \text{ M}} \times 25 \text{ ml} = 10 \text{ ml of } MgSO_4 \text{ stock}$$

Step 3. Q.S. to 25 ml:

25 ml – (2.5 ml + 10 ml) = 12.5 ml of water

It is always a good idea to check your work. Use the method for calculating concentrations from recipes given on the following pages.

If you are interested in understanding this better, you can think of it in terms of proportions. The proportion of the final volume that is contributed by the concentrated solution

$$\frac{\text{Volume to add}}{\text{Final volume}}$$

should be the same as the proportion of the final concentration of chemical that is contributed by the concentrated solution:

$$\frac{\text{What you want}}{\text{What you have}}$$

That is:

$$\frac{\text{Volume to add}}{\text{Final volume}} = \frac{\text{What you want}}{\text{What you have}}$$

Now, rearrange and you have the How To Quickly.

HOW TO QUICKLY Dilute to a Particular Concentration
Plug your numbers into the following word equation:

$$\frac{\text{What you want}}{\text{What you have}} \times \text{Final volume} = \text{Volume to add to mixture}$$

Q.S. to the final volume.

What you want = Desired final concentration [M]
What you have = Concentration of your stock solution [M]
Final volume = Total volume of solution with final concentration [liter]
Volume to add to mixture = Volume of stock solution to be diluted [liter]

"What you want" and "what you have" must have the same units.

See also Plug and Chug in Chapter 8.

Using Dilution Ratios

Another common method for figuring out how to dilute a solution to a particular concentration is to determine the ratio of diluted to undiluted (stock) concentrations, and then convert that ratio to a fraction and simplify: That fraction of the final volume should be made of concentrated solution; the rest should be solvent.

Revisiting the above example, you want to make 25 ml of a solution that is 0.5 M $CaCl_2$, 1.0 M $MgSO_4$. The stock solutions available are 5.0 M $CaCl_2$ and 2.5 M $MgSO_4$. For $CaCl_2$, the ratio of diluted to undiluted is 0.5:5.0, which simplifies to 1:10; therefore, the 5.0 M $CaCl_2$ should make up 1/10th of the final volume. The final volume is 25 ml, 1/10th x 25 ml = 2.5 ml; therefore, you should add 2.5 ml of 5.0 M $CaCl_2$. For $MgSO_4$, the ratio is 1:2.5; therefore, the 1.0 M $MgSO_4$ should make up 1/2.5th times the final volume, which equals 10 ml. As before, 10 + 2.5 = 12.5, and 25 − 12.5 = 12.5, so it will take 12.5 ml of solvent to Q.S. to the final volume of 25 ml. Written as a word equation, this looks like:

$$\frac{[\text{Diluted}]}{[\text{Undiluted}]} \times \text{Final volume} = \text{Volume to add}$$

[Diluted] = Concentration of chemical in the final diluted solution [M]
[Undiluted] = Concentration of chemical in the starting stock solution [M]
Final volume = Volume of diluted solution needed
Volume to add = Volume of undiluted stock solution to use

HOW TO QUICKLY Use Ratios to Dilute to a Particular Concentration

1. Write the ratio of diluted concentration to undiluted concentration as a fraction.

2. That fraction times the total volume is the volume of undiluted stock solution to add.

3. Repeat for each stock solution to be diluted and used.

4. Q.S. to the final volume.

Diluted concentration = Concentration you are working to achieve [M]
Undiluted concentration = Concentration of the solution you will be diluting [M]
Final volume = Volume of diluted solution that you want to make

MAKING SERIAL DILUTIONS

If you start with a concentrated solution and dilute it, then take some of that diluted stock and dilute it more, then take some of that diluted diluted stock and dilute it some more, then take some of that diluted, diluted, diluted stock and dilute it some more, etc., you are doing a serial dilution. A series of dilutions like this is a quick and easy way to produce a solution in a variety of concentrations and is also sometimes the only way to effectively and accurately dilute an extremely concentrated solution or, often, cell suspension.

   Often, the goal of a serial dilution is to make a variety of concentrations so that you can then determine which concentration is most appropriate for the job at hand. You could make each of your dilutions separately using:

$$\frac{\text{Concentration you want}}{\text{Concentration you have}} \times \text{Final volume} = \text{Volume to add}$$

But if you want a series of different dilutions that vary in a constant way, or if the dilution you want is *very* dilute but you want to conserve solvent, or if you can't measure out a small enough volume of what you have to dilute it accurately, it is easier to start with the high concentration and do serial dilutions. The difference between each of the various concentrations that result from a serial dilution is constant. For example, each is one-half the concentration of the next highest, or each is one-tenth the concentration of the next highest.

Serial dilutions are often used to dilute cell suspensions. In this case, the units of concentration are cells/ml; it works the same way. Serial dilutions are particularly well-suited for cell suspensions because these suspensions are often too concentrated to dilute easily in one step.

METHOD **Serial Dilution the Simple, but Imprecise Way**

If you don't have to worry about saving reagents or minimizing waste, you can do certain serial dilutions very easily. Specifically, 1:10 and 1:2 dilutions are a snap.

1:10 Serial Dilutions

1. Figure out how many different dilutions you want, and about how much volume of each you need. Make about 20% more volume than you need; use containers large enough to hold about 50% more than the volume you are making. Put the containers in a row on the bench, and label them 1:10, 1:100, 1:1000, etc., or even better, label them with the concentrations of the solutions they will hold.

2. Put 9/10ths of the volume-you-are-making (i.e., 9/10ths the volume you need plus 20%) of solvent in each container.

3. Take 1/10th of the volume-you-are-making of concentrated solution or suspension, and dispense it into the 1:10 container. Mix.

4. Take 1/10th of the volume-you-are-making of 1:10 solution or suspension, and dispense it into the 1:100 container. Mix.

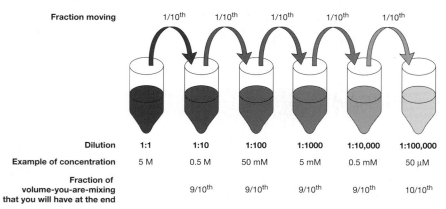

Fraction moving		$1/10^{th}$	$1/10^{th}$	$1/10^{th}$	$1/10^{th}$	$1/10^{th}$
Dilution	1:1	1:10	1:100	1:1000	1:10,000	1:100,000
Example of concentration	5 M	0.5 M	50 mM	5 mM	0.5 mM	50 µM
Fraction of volume-you-are-mixing that you will have at the end		$9/10^{th}$	$9/10^{th}$	$9/10^{th}$	$9/10^{th}$	$10/10^{th}$

5. Take 1/10th of the volume-you-are-making of 1:100 solution or suspension, and dispense it into the 1:1000 container. Mix.

6. And so on.

1:2 Serial Dilutions

1. Figure out how many different dilutions you want, and about how much volume of each you need. Make about double the volume that you need; use containers large enough to hold more than double the volume you are making. Put the containers in a row on the bench, and label them 1:2, 1:4, 1:8, etc., or even better, label them with the concentrations of the solutions they will hold.

2. Put the volume-you-are-making (i.e., double the volume you need) of solvent in each container.

3. Take the volume-you-are-making of concentrated solution or suspension, and dispense it into the 1:2 container. Mix.

4. Take the volume-you-are-making of 1:2 solution or suspension, and dispense it into the 1:4 container. Mix.

5. Take the volume-you-are-making of 1:4 solution or suspension, and dispense it into the 1:8 container. Mix.

6. And so on.

| METHOD | Serial Dilution the Slightly Less Simple but More Precise Way |

If you need to be very precise with your reagents because they are either expensive or hazardous, the following steps will tell you how to make exactly the amount you need.

1. Decide what final volume you need (V_f) of each dilution. (If you need to dilute to a particular concentration, see page 115).

2. Decide how many different concentrations (N) you need, and line up that many containers.

3. Decide what the dilution factor (X) will be. If you are doing a first experiment to determine the appropriate concentration, a good rule of thumb is to cover a few orders of magnitude, meaning that you

will be doing 1:10 dilutions. If you are doing an experiment to narrow down the concentration, do 1:2 dilutions.

4. Calculate what the concentration will be in each container when you are done:

   - The first will have a concentration of $\frac{1}{X}$ × initial concentration (C_i), where X is the dilution factor.

   - The second will have a concentration of $\frac{1}{X^2}$ × initial concentration (C_i).

   - The third will have a concentration of $\frac{1}{X^3}$ × initial concentration (C_i).

   - And so on.

5. Label the containers with the final concentration they will hold.

6. Put V_f solvent into each container.

7. Set your pipette to $\frac{V_f}{X-1}$.

8. Go:

   Put a new pipette tip on your pipetter, or get a new pipette.

   Pick up $\frac{V_f}{X-1}$ of the most concentrated solution.

   Dispense it into the first vessel, the one whose label indicates that the concentration will be

   $\frac{1}{X}$ × initial concentration

   Mix.

   Change pipette tip or pipette.

9. Pick up $\frac{V_f}{X-1}$ of the solution you just made. (You will be leaving behind exactly the volume you need of that dilution.)

   Dispense it into the next vessel, the one whose label indicates that the concentration will be

   $\frac{1}{X^2}$ × initial concentration

   Mix.

   Change pipette tip or pipette.

10. Pick up $\dfrac{V_f}{X-1}$ of the solution you just made.

Dispense it into the next vessel, the one whose label indicates that the concentration will be

$\dfrac{1}{X^3}$ x initial concentration

Mix.

Change pipette tip or pipette.

11. And so on.

☑ **Example 1**

For a 1:10 serial dilution of a 10 M (C_i) solution, you need:

10 ml of each dilution (V_f = 10 ml)
6 different concentrations (N = 6)
1:10 dilutions (X = 10)

Note: 1:10 dilutions are the simplest to calculate, because each dilution is 1/10th the concentration of the last, which means, whatever the concentration was, just move the decimal point one digit to the left to get the concentration of the next most dilute solution.

Container 1: $\dfrac{10\ M}{10}$ = 1 M

Container 2: $\dfrac{10\ M}{10^2}$ = 0.1 M (or $\dfrac{1\ M}{10}$ = 0.1 M)

Container 3: $\dfrac{10\ M}{10^3}$ = 0.01 M (or $\dfrac{0.1\ M}{10}$ = 0.01 M)

Container 4: $\dfrac{10\ M}{10^4}$ = 0.001 M (or $\dfrac{0.01\ M}{10}$ = 0.001 M) = 10^{-3} M = 1 mM

Container 5: $\dfrac{10\ M}{10^5}$ = 0.0001 M (or $\dfrac{0.001\ M}{10}$ = 0.0001 M) = 10^{-4} M

Container 6: $\dfrac{10\ M}{10^6}$ = 0.00001 M (or $\dfrac{0.0001\ M}{10}$ = 0.00001 M) = 10^{-5} M = 10 μM

1. Label the containers

#1:   1 M
#2:   0.1 M (= 1 x 10^{-1} M)
#3:   0.01 M (= 1 x 10^{-2} M)

#4:   0.001 M (= 1×10^{-3} M = 1 mM)
#5: 1×10^{-4} M
#6: 1×10^{-5} M (= 10 μM)

2. Put 10 ml (V_f) of solvent into each container.

3. Set your pipette to $\dfrac{V_f}{X-1} = \dfrac{10 \text{ ml}}{10-1} = \dfrac{10 \text{ ml}}{9} = 1.111$ ml.

4. Go:

   Put a new pipette tip on your pipetter.

   Pick up 1.111 ml of the 10 M solution.

   Dispense it into the container labeled #1: 1 M.

   Mix.

   Change the pipette tip.

5. Pick up 1.111 ml of the #1: 1 M solution.

   Dispense it into the container labeled #2: 1×10^{-1} M.

   Mix.

   Change the pipette tip.

6. Pick up 1.111 ml of the #2: 1×10^{-1} M solution.

   Dispense it into the container labeled #3: 1×10^{-2} M.

   Mix.

   Change the pipette tip.

7. Pick up 1.111 ml of the #3: 1×10^{-2} M solution.

   Dispense it into the container labeled #4: 1×10^{-3} M (= 1 mM).

   Mix.

   Change the pipette tip.

8. Pick up 1.111 ml of the #4: 1×10^{-3} M solution.

   Dispense it into the container labeled #5: 1×10^{-4} M.

   Mix.

   Change the pipette tip.

9. Pick up 1.111 ml of the #5: 1×10^{-4} M solution.

   Dispense it into the container labeled #6: 1×10^{-5} M (= 10 μM).

   Mix.

   Throw away the last pipette tip.

☑ Example 2

For 1:2 serial dilution of a 12.6 mg/ml solution, you need:

150 ml of each dilution
3 different concentrations
$X = 2$

Container 1: $\dfrac{12.6 \text{ mg/ml}}{2}$ = 6.30 mg/ml

Container 2: $\dfrac{12.6 \text{ mg/ml}}{2^2}$ = 3.15 mg/ml (or $\dfrac{6.30 \text{ mg/ml}}{2}$ = 3.15 mg/ml)

Container 3: $\dfrac{12.6 \text{ mg/ml}}{2^3}$ = 1.57 mg/ml (or $\dfrac{3.15 \text{ mg/ml}}{2}$ = 1.57 mg/ml)

1. Label the containers:

   #1: 6.30 mg/ml
   #2: 3.15 mg/ml
   #3: 1.57 mg/ml

2. Put 150 ml of solvent into each container.

3. Pick a graduated cylinder that holds $\dfrac{150 \text{ ml}}{2-1}$ = 150 ml.

4. Go:

   Measure out 150 ml of 12.6 mg/ml solution.

   Dispense it into the container labeled #1: 6.30 mg/ml.

   Mix.

   Clean the cylinder or get an unused one.

5. Measure out 150 ml of #1: 6.30 mg/ml solution.

   Dispense it into the container labeled #2: 3.15 mg/ml.

   Mix.

   Clean the cylinder or get an unused one.

6. Measure out 150 ml of #2: 3.15 mg/ml solution.

Dispense it into the container labeled #3: 1.57 mg/ml.

Mix.

Wash the glassware.

This example illustrates that when picking out your containers, you need to account for what the volume will be after the more-concentrated solution has been pipetted in, and before the less-concentrated solution has been pipetted out. In other words, your container must hold $V_f + \frac{V_f}{X-1}$. If you are doing 1:2 dilutions, you must pick out a container that can hold twice the final volume. You will always end up with $V_f + \frac{V_f}{X-1}$ of the final dilution—another reason to use a large enough container

In brief, do the calculations for your dilutions first, and then set up a series of *carefully labeled* containers, in order. Add the appropriate amount of solvent to the containers, and then pipette concentrate into the first container, mix, change pipette tips, pipette the newly diluted solution into the next container, mix, change pipette tips, repeat. The mixing and changing of pipette tips is not just cosmetic: If you don't do it, your concentrations will not be correct. The following isn't exactly a How To Quickly, but it does summarize the process.

How to Do a Serial Dilution

Set up N containers with capacities to hold $V_f + \dfrac{V_f}{X-1}$.

Label them $\dfrac{C_i}{X}, \dfrac{C_i}{X^2}, \dfrac{C_i}{X^3}, ..., \dfrac{C_i}{X^N}$

Put V_f solvent into each container.

Put $\dfrac{V_f}{X-1}$ of solution C_i into container $\dfrac{C_i}{X}$. Mix. Change pipette.

Put $\dfrac{V_f}{X-1}$ of solution $\dfrac{C_i}{X}$ into container $\dfrac{C_i}{X^2}$. Mix. Change pipette.

Put $\dfrac{V_f}{X-1}$ of solution $\dfrac{C_i}{X^2}$ into container $\dfrac{C_i}{X^3}$. Mix. Change pipette.

Put $\dfrac{V_f}{X-1}$ of solution $\dfrac{C_i}{X^{N-1}}$ into container $\dfrac{C_i}{X^N}$. Mix. Toss pipette.

N = Number of different concentrations
V_f = Final volume of each diluted solution
X = Dilution factor
C_i = Initial concentration [M]

See also Plug and Chug in Chapter 8.

Serial Dilutions to Achieve a Particular Concentration

Sometimes you need to serially dilute a solution or a cell suspension to a particular concentration; in this case, you aren't interested in the concentrations along the way, only the final concentration. A common scenario is diluting a cell suspension so that you can replate the cells at an optimal density.

First, you have to figure out the final dilution. If that dilution is obviously a power of an integer, for example, 1:10,000 and 10,000 is 10^4, you can do four 1:10 dilutions (or two 1:100 dilutions, or 1:10 once and 1:1000 once; the sum of the exponents on the tens must equal 4 because 4 was the exponent on the 10 of the originally calculated final dilution). More likely, each dilution will be different.

An easy way to plan each dilution is to calculate the final dilution and then write that number in scientific notation. The first dilution should be 1:the mantissa; the rest will be 1:10 (or 1:100, etc.) dilutions. So, if you calculated that the final dilution would be 1:36,850, that's 3.6850×10^4. The first dilution would therefore be 1:3.685, for example, 1 ml of suspension plus 2.685 ml of medium. Then dilute 1:10 four times (or 1:100 two times, as above).

☑ Example

You have *Escherichia coli* suspended at 3×10^6 cells/ml, which is 3×10^3 cells/µl. You want to be able to replate so that you will end up with about 10 colonies, and you need those 10 cells suspended in about 50 µl for delivery. So, the final concentration you need is 10 cells/50 µl, which equals 0.2 cells/µl. You want to make 6 new plates, so you will need a total of 300 µl of this final suspension. Your best pipetter can pick up 1 µl accurately.

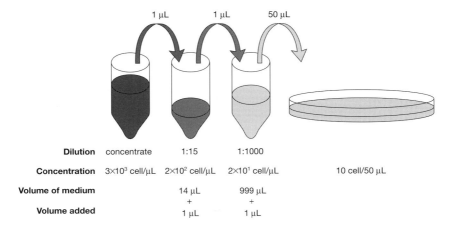

Dilution	concentrate	1:15	1:1000	
Concentration	3×10^3 cell/µL	2×10^2 cell/µL	2×10^1 cell/µL	10 cell/50 µL
Volume of medium		14 µL	999 µL	
		+	+	
Volume added		1 µL	1 µL	

The ratio of diluted to undiluted is $0.2{:}3 \times 10^3$ or, $1{:}15 \times 10^3$. So, the first dilution can be 1:15 (1 μl of concentrated suspension + 14 μl of medium), and the second, 1:1000 (1 μl of the first dilution + 999 μl of medium).

You could also have done 1:150 for the first dilution (1 μl of concentrate + 149 μl of medium) and 1:100 for the second (careful: you need to end up with 300 μl, so the second dilution would have to be 3 μl of the first dilution + 297 μl of medium).

There is no difference in what you end up with (there is a difference in the amount of waste generated: 714 μl in the first case, 147 μl in the second). If you have a dilution you do frequently, you can fiddle with the numbers and determine the optimal (least wasteful or easiest) series. If you do that, write it down for everyone.

CONVERTING RECIPES TO CONCENTRATIONS

Sometimes, when you are reading scientific literature or someone's protocol for making a solution or performing an experiment, you will encounter a recipe for exactly how to make a particular volume of a particular solution (e.g., dissolve 2.7 mg of EDTA in 40 ml of water). You, however, must know the final concentration of the EDTA in solution, not simply how to make up the same solution. If this happens, you have to know how to convert a recipe into a concentration.

Essentially, this is a word problem, and the trick to solving a word problem is to convert the words into math. The key to this process is knowing the translation. The translations you need to make solutions are:

- "Of" means multiply.

- "Put amount *A* into volume *B*" (or add *B* volume to *A* amount) means *A* divided by *B*.

It can help to start by drawing a picture of what was done. This may seem like a time waster, but before you dismiss the idea, think about the amount of time you would otherwise spend scribbling random equations on scrap paper.

☑ Example

Suppose the Materials and Methods section said "100 μl of D-(+)-glucose (ChemCo, 1 mg/ml) was diluted 1:100 in DME." That means that the following actually happened:

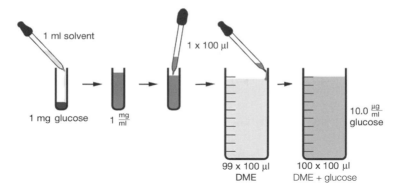

Start by translating what was written into a description of what was done; use the words "of" and "put into." 1 mg OF glucose was PUT INTO 1 ml of solvent (ChemCo did that), or:

1 mg ÷ 1 ml = 1 mg/ml

100 µl OF 1 mg/ml glucose solution was PUT INTO a total of 10 ml of solution, or:

$$\frac{100 \text{ µl} \times 1 \text{ mg/ml}}{10 \text{ ml}}$$

Calculate:

$$\frac{100 \text{ µl} \times 1 \text{ mg/ml} \times 10^{-3} \text{ ml/µl}}{10 \text{ ml}} = 10^{-2} \text{ mg/ml} = 10 \text{ µg/ml}$$

So, the concentration they used was:

10 µg/ml glucose in DME

pH

pH is the way that the acidity of a solution is quantified. The pH of a solution is usually important and often critical, since many reactions will only take place as expected if the pH is correct. Sometimes, pH is essential to the solution-making process itself. For example, EDTA will not dissolve until the pH is brought to 7.

In the term pH, the "p" means "the negative of the log of"; the "H" stands for "the concentration of protons." If pH = 6, that means $[H^+]$ =

10^{-6}; if pH = 8, that means [H$^+$] = 10^{-8}. So, pH is a logarithmic measure of the concentration of protons in a solution. Because pH is the *negative* of log [H$^+$], the *lower* the pH, the *higher* the concentration of H$^+$, and the *higher* the pH, the *lower* the concentration of H$^+$. The pH scale runs from 1 to 14, with pH 7 (the pH of pure water) considered neutral. A pH below 7 means a solution is acidic; a pH above 7 means a solution is basic (or alkaline).

Because pH is logarithmic, a change in pH of 1 means a 10^1x or a 10x change in [H$^+$]. A solution of pH 6, therefore, has 10 times as many protons as a solution of pH 7. A change in pH of 2 means a 10^2x or 100x change in [H$^+$]; a change in pH of 3 means a 10^3x or 1000x change in [H$^+$]; etc. pH is not typically calculated; it is usually measured directly using a pH meter or estimated using pH paper.

Measuring pH

To measure pH, you can use indicator dyes, litmus paper, or a pH meter. A solution being "pH'd" must be well-stirred.

Indicator dyes

Indicator dyes change color when pH changes. These dyes are calibrated so that you can compare by eye the color of a solution with dye in it to a chart with a key of colors and their corresponding pHs. Indicator dyes are the pH-measuring method most pet stores sell for measuring the pH of fish tanks. Many biological media contain indicator dyes, like phenol red, so that users can tell at a glance whether the pH of their solution is correct.

Litmus paper

Litmus paper is coated with an indicator dye that changes color with pH. It is calibrated so that you can easily observe the color that results when you drop some of your solution onto the paper (never dip the paper into the solution—you will contaminate it). By comparing the color to the chart provided, you can determine the pH. Litmus paper comes in a variety of pH ranges to accommodate different precisions of measurement.

pH meters

pH meters work by measuring voltage changes that occur when an electrode is placed in a solution. The tip of the electrode is permeable to hydrogen ions. When the electrode is placed in a solution, a certain number of ions, proportional to the pH, get inside the electrode and change the electrical properties of a sensor. That change is monitored and converted into a measure of pH.

The conversion made by a pH meter is dependent on calibration, which is why it is critical to calibrate any pH meter properly before each use. For more information about pH meters, see Chapter 3. pH meters should be calibrated using at least two buffers that bracket the pH being measured. If at any time you are uncertain in any way about your pH meter, use caution and re-calibrate it.

> **Note:** There are two very important things to remember when using a pH meter. One is that the electrode of many pH meters must not dry out. The electrode should always be kept submerged in liquid, even when the meter is not in use. You can buy storage buffers, or just store the electrode in one of the standard buffers (pH 7 or pH 4). The second thing is never to transfer the electrode from one liquid to another without first rinsing it well, for example, with distilled water (catch the runoff in a waste container).

METHOD Measuring pH with a pH Meter

Below are some general guidelines. Follow the precise instructions that came with your pH meter to calibrate and use your particular meter properly.

1. Calibrate the pH meter.

   a. To calibrate a pH meter, choose buffers you will standardize to that bracket the final pH you are measuring. For most biological applications, using standards of pH 4.0 and of pH 10 will bracket the pHs you are likely to measure. Another common standard to use is pH 7. If you will only be measuring pHs below 7, calibrate using standards of pH 4 and pH 7; if you will only be measuring pHs above 7, use standards of pH 7 and pH 10. Some pH meters allow you to use all three standards to calibrate. Whichever standards you use, the meter is taking the calibration points and graphing a line that will tell it the relationship of voltage to pH at that moment in time. You must use at least two standards, because it takes two points to define a line. The instructions that come with your meter and the electrode will have more details about calibrating and topics such as the uncertainty associated with measuring extreme pHs.

   b. Submerge the *well-rinsed* electrode into the pH 4.0 standard. Use the appropriate button or knob to set the pH to 4.0.

c. Submerge the *well-rinsed* electrode into the pH 10 standard. Use the appropriate button or knob to set the pH to 10.

2. Put the *well-rinsed* electrode tip into the well-mixed (usually contemporaneously mixing) solution to be measured.

3. Wait for the measurement of pH to stop fluctuating.

4. Read the pH.

From Acid-Base to Buffers: Conjugate Acid-Base Pairs

Every acid has a conjugate base, and every base has a conjugate acid. An acid's conjugate base has one fewer H and one more negative charge than the acid; a base's conjugate acid has one more H and one less negative charge than the base. In other words, if an acid gives up a proton, what's left is its conjugate base; if a base picks up a proton, the result is its conjugate acid.

☑ Example

$$H_2PO_4^- \leftrightarrow H^+ + HPO_4^{2-}$$
Acid   Conjugate base

A buffer is a weak acid that is chosen such that it prevents changes in pH by substituting changes in the relative concentrations of the weak acid and its conjugate base. So a buffer works by replacing a change in $[H^+]$ with a change in relative amounts of the weak acid and its conjugate base in the solution. When you add acid to a buffered solution, the pH stays the same, the amount of buffer goes up, and the amount of the buffer's conjugate base goes down. When you add base to a buffered solution, the pH stays the same, the amount of buffer goes down, and the amount of its conjugate base goes up.

☑ Examples

Compare the following scenarios:

1. You add NaOH to a solution:

   Result: The NaOH dissociates into Na^+ and OH^-. The pH of the solution consequently rises. OH^- provided by the dissociation of the NaOH combines with free H^+, bringing down the $[H^+]$ in the solution. In addition, the number of molecules of Na^+ goes up, as does the number of molecules of H_2O.

2. You add NaOH to a buffered solution:

> **Result:** The NaOH dissociates into Na^+ and OH^-. The acid component *of the buffer* donates H^+ to combine with the free OH^- and the pH does not change. What *does* change is the relative amount of acid (decreases) and conjugate base (increases) in the solution.

pK_a: Judging a buffer by its number

An important number used to characterize buffers is pK_a. A full explanation of pK_a is given below, but the simple, critical thing to remember about pK_a and buffers is that you want to choose a buffer with a pK_a as close as possible to the pH you ultimately want to maintain. For biological solutions, a good rule of thumb is to get as close as you can or a little lower. Although it is not covered in this book (because readily available buffers work quite well for most purposes), you might want to know that there is a way to design and make your own buffers with particular pK_as. If you do need a buffer that doesn't seem to exist, ask a friendly biochemist for help using the Henderson-Hasselbalch equation (see below) to create a recipe for the buffer you need.

Acid-base reactions and pK_a

To understand pK_a, the number that describes buffers, you can think about the classical definitions of acids and bases. Every acid-base reaction resembles this generic formula:

$$HA + H_2O \Leftrightarrow H_3O^+ + A^-$$

which, because the H_2O to H_3O^+ is assumed, is sometimes written:

$$HA \Leftrightarrow H^+ + A^-$$

where HA is an acid and H_2O is the base it reacts with to form H_3O^+ (or H^+, the conjugate acid of H_2O), and A^- (the conjugate base of HA). This generic reaction is quantified by the dissociation constant, which is the acid-base equivalent of an equilibrium constant (K) (for more on equilibrium constants, see page 94):

$$K = \frac{[H_3O][A^-]}{[HA][H_2O]}$$

K is (in this case) a dimensionless quantity that measures the proton (i.e., H^+) affinity of the HA/A^- pair relative to the proton affinity of the

H_3O^+/H_2O pair; that is, it tells you whether you are more likely to have HA or H_3O^+. In other words, K tells you whether the generic reaction (HA $\Leftrightarrow$ H^+ + A^-) is more likely to go to the right (the acid is strong and gives up a proton easily; the acid will dissociate even if there are already lots of protons around) or to the left (the acid is weak and does not give up a proton easily; the acid will not dissociate so easily). If K is above 1.0, the reaction goes to the right and HA is a stronger acid (one that readily dissociates). If K is below 1.0, the reaction goes to the left, and HA is a weaker acid (one that doesn't dissociate easily). To get from K to K_a, just rearrange the definition of K:

$$K[H_2O] = \frac{[H_3O^+][A^-]}{[HA]} = K_a$$

and that is the definition of K_a:

$$K_a = \frac{[H_3O^+][A^-]}{[HA]} = \frac{[H^+][A^-]}{[HA]}$$

Careful: Sometimes the subscript is left off.

Interesting aside: Why pH 7, the pH of pure water, is neutral

H_2O is an acid (albeit a weak one), so it has a dissociation constant:

$$K_a = \frac{[H^+][OH^-]}{[H_2O]}$$

Rearranging gives:

$$K_a [H_2O] = [H^+][OH^-]$$

and that is the definition of K_w, the dissociation constant for water.

$$K_w = K_a[H_2O] = [H^+][OH^-]$$

At 25°C, $K_w = 10^{-14}$ M^2. Moreover, because in pure water the concentration of H^+ must equal the concentration of OH^- (because they are present in equimolar amounts in H_2O), $[H^+]$ must equal the square root of K_w, that is, 10^{-7} M. If $[H^+] = 10^{-7}$ M, then pH = 7. Hence, the pH of pure water is 7, and that is defined as neutral.

The Relationship of pK_a to pH

We can manipulate the definition of K_a to show the relationship between pH and the concentration of an acid and its conjugate base. First, taking the log of both sides of the equation defining K_a gives:

$$\log K_a = \log [H^+] + \log [A^-] - \log [HA]$$

Rearranging gives:

$$-\log [H^+] = -\log K_a + \log [A^-] - \log [HA]$$

Which is the same as:

$$-\log [H^+] = -\log K_a + \log \frac{[A^-]}{[HA]}$$

Now, we substitute the shorthand "p" for "–log" and get:

$$pH = pK_a + \log \frac{[A^-]}{[HA]}$$

This is the Henderson-Hasselbalch equation, and it tells you the relationship between the concentration of an acid [HA], the concentration of its conjugate base [A$^-$], and pH. The Henderson-Hasselbalch equation is useful because it allows you to talk about the strength of an acid (the readiness with which it dissociates) with reference to the pH scale. To say an acid is strong is to say that it dissociates (donates protons), even if there is already a high concentration of protons around (i.e., it has a high K_a, which means it has a low pK_a; rule of thumb: $K_a > 10^{-2}$ is a large K_a, so a pK_a of 2 is low). If the pK_a of an acid is 2, then in a solution of pH = 2, [A$^-$] must equal [HA] (because the second term on the right-hand side must equal zero; so the fraction must equal one), which means that half of the acid has dissociated. Any chemical that can be 50% dissociated into protons and conjugate base in a solution that already has a pH of 2 is certainly a strong acid. If the pK_a of another acid is 5, then in a solution of pH 2, the second term must equal –3; so the fraction must equal 10^{-3} which means [A$^-$] is a thousand times lower than [HA], which means less of this second acid dissociated, i.e., it is a weaker acid.

Let's go back to the Henderson-Hasselbalch equation, which says that if [A$^-$] = [HA] then pH = pK_a. This relationship between pH and pK_a is important to understand: When choosing an acid to use as a buffer, you *want* [HA] to be about equal to [A$^-$] because that will be an acid that is about equally good at buffering against increases or decreases in pH. This means you want a buffer with a pK_a close to the pH you wish to maintain. Because buffers only work over a certain range of pHs (rule of thumb: buffers buffer in a range of $pK_a \pm 1$), you want to get as close as possible.

HOW TO QUICKLY Choose a Buffer

1. Decide what pH you would like to maintain in your solution.

2. Look up buffers in the table in Chapter 8 or in a catalog.

3. Choose a buffer with a pK_a close to the desired pH.

4. Keep in mind that the buffer will only work over a certain range of pHs ($pK_a \pm 1$).

Warning: pH is sensitive to changes in temperature and to what is in the solution being buffered. Also, different buffers work differently. Unless you are absolutely positive, due to long experience with a particular buffered solution, it's a very good idea to confirm the pH of important solutions close to the time you will be using them.

Biological solutions frequently become acidic over time. When selecting a buffer to use in a biological application, choose one with a pK_a that is slightly lower than the desired pH. That way, as the pH of the solution tries to change, the buffer is still working over the range at which it is good at buffering.

Some buffers can donate more than one proton and thus have more than one pK_a. The range over which these buffers work will therefore be larger than the range of buffering possible from an acid that donates only one proton.

How much buffer to include in a solution

Most recipes will tell you how much buffer to add. If you need to figure it out for yourself, you should know that there are no hard-and-fast rules for deciding how much buffer to add, but there are some things to consider. The concentrations of various buffers in humans are as follows:

$[HCO_3^-]$ = 24 mM
$[HPO_4^{2-}]$ = 1 mM
$[SO_4^{2-}]$ = 0.5 mM

For solutions to be used in biological applications, buffer concentrations are usually between 10 mM and 100 mM. Some catalogs include a good working concentration for the buffers they sell. The higher the concentration of buffer, the more you have to worry about the effect the buffer may have on the ionic strength and/or the osmolarity of your solution.

METHOD Adjusting pH Using a pH Meter

1. Calibrate the meter.

   a. Choose buffers you will standardize to that bracket the pH you are aiming for. For most biological applications, using a standard of pH 4.0 and a standard of pH 10 will bracket the pHs you are likely to measure.

   b. Submerge the *well-rinsed* electrode into the pH 4.0 standard. Use the appropriate button or knob to set the pH to 4.0.

   c. Submerge the *well-rinsed* electrode into the pH 10 standard. Use the appropriate button or knob to set the pH to 10.

2. Put the *well-rinsed* electrode tip into the actively mixing solution.

3. Wait for the measure of pH to stop fluctuating.

4. Read the pH.

5. Change the pH.

   a. If the pH of a solution is too low:

   Checking the pH between the addition of *each* drop, add *one drop at a time* of high-concentration (usually 10 N) NaOH (or KOH) until the pH is within 1.0 of the desired value. (Remember, the pH is a log scale, whereas the number of H^+ or OH^- molecules in the drops you are adding remains constant. So, one drop will affect a solution with a pH of 5 ($[H^+] = 10^{-5}$ M) much less than it will affect a solution with a pH of 10 ($[H^+] = 10^{-10}$ M). Then, *one drop at a time*, add low-concentration (1.0 N) NaOH (or KOH).

   b. If the pH of a solution is too high:

   Checking the pH between the addition of *each* drop, add *one drop at a time* of high-concentration (usually 12.1 N) HCl until the pH is within 1.0 of the desired value. Then, *one drop at a time*, add low-concentration (usually 1.2 N) HCl.

RESOURCES

Safety

http://hazard.com/msds/

http://www.cdc.gov/niosh/ipcs/icstart.html

http://www.ilpi.com/msds/index.html

http://www-sci.lib.uci.edu/HSG/GradBioscience.html

Vocabulary

http://www.netaccess.on.ca/~dbc/cic_hamilton/chemed.html

www.chem.purdue.edu/gchelp/gloss/terms.html

http://www.lhup.edu/~dsimanek/glossary.htm

Lab Techniques

Barker K. 1998. *At the bench: A laboratory navigator.* Cold Spring Harbor Laboratory Press, Cold Spring Harbor, New York.

http://www.dartmouth.edu/~chemlab

Acid-base and Buffers

http://www.sbu.ac.uk/biology/biolchem/acids.html

http://www.albany.edu/faculty/dab/1-24.html

Henderson-Hasselbalch Equation

www.sbu.ac.uk/biology/biolchem/acids.html

CHAPTER 5

DNA and RNA

TALKING ABOUT DNA AND RNA

Absorbance: The property of a substance that describes its ability to interact with and redirect photons. The ability of certain molecules to absorb and re-emit photons is the basis of spectrophotometry, in which light of a known intensity is beamed through a solution, and the intensity of the light emerging on the far side unaffected is measured. The amount of light "lost" inside the solution is directly proportional to the concentration of absorbing molecules; thus, absorbance can be used to measure concentration.

Base: The purines (adenine and guanine) and pyrimidines (cytosine, thymine, and uracil); the nitrogen-containing ring compounds that, when coupled by a sugar phosphate backbone, make up the nucleic acids DNA and RNA. The bases in DNA are adenine (A), cytosine (C), thymine (T), and guanine (G). In RNA, uracil (U) appears instead of thymine. The sequence of the bases constitutes the information-coding ability of the nucleic acids; the ability of A to pair with T (U) and G to pair with C is the basis of the replication of nucleic acids and of transcription.

Base Codes: A, C, T, G, and U are the symbols used to indicate bases or nucleotides. There are also symbols that mean either of two bases/ nucleotides, any of three bases/nucleotides, or any base/nucleotide at all.

Code	What It Represents	Code	What It Represents
A	Adenine	S	C or G
C	Cytosine	Y	C or T
G	Guanine	K	G or T
T	Thymine	V	A, C, or G (not T)
U	Uracil	H	A, C, or T (not G)
M	A or C	D	A, G, or T (not C)
R	A or G	B	C, G, or T (not A)
W	A or T	N	A, C, G, or T

bp: Abbreviation for "base pair" or two nucleotides in DNA (or RNA) paired by a hydrogen bond. A pairs with T in DNA, A pairs with U in RNA, and G pairs with C. See also Base.

Dalton: Unit of molecular mass, equal to *amu*. One molecule of dAMP has a mass of 313.2 daltons (D).

ds: Abbreviation for "double-stranded" applied to DNA and RNA.

kb: Abbreviation for kilobase, i.e., 1000 bases; 1000 base pairs can be assumed if the reference is to a double-stranded nucleic acid.

Molecular Weight (MW; also called the Molecular Molar Mass): This refers to the mass of the molecular formula. The molecular weight of neomycin sulfate (formula: $C_{23}H_{46}N_6O_{13} \cdot 3H_2SO_4$) is 908.9. The units of molecular weight are g mol^{-1}(if describing moles of the molecule) or D (if describing individual molecules). Often, these units are not written out. This term is now sharing space with "Relative Molecular Weight" (M_r), which is dimensionless.

Relative Molecular Weight (M_r): This is the molecular weight of a molecule expressed as a ratio of the mass of the molecule to the mass of 1/12th of a ^{12}C atom; 1/12th of a ^{12}C atom has a mass of 1 D, and so the magnitude of the number does not change. So, why bother? Because M_r is a ratio, it is dimensionless. In practice, because the magnitude of the number is the same as the molecular weight, there can be confusion. If you are reporting the molecular weight of a protein, you should write MW = 10 kD or 10,000 D and refer to it as the 10-kD protein. If you are reporting the relative molecular weight, you should write M_r = 10,000 (no units because M_r is dimensionless) and refer to it as the 10,000 M_r protein.

Nucleic Acid: DNA and RNA; polymers composed of nucleotide subunits covalently linked by phosphodiester bonds. The structure is a sugar-phosphate backbone (technically, a phosphate-pentose polymer) with protruding bases (technically, bases as side groups); two antiparallel strands linked by hydrogen bonds between complementary bases form a double helix. At physiological pH, the phosphates give up protons; hence, nucleic acid. The net charge on a nucleic acid is negative.

Nucleoside: A base plus a deoxyribose or ribose (a 5-carbon sugar); a nucleotide minus the phosphates. The nucleosides are deoxyadenosine, deoxyguanosine, deoxycytidine, deoxythymidine, adenosine, guanosine, cytidine, and uridine.

Nucleotide: A base, plus a deoxyribose or a ribose, plus one or more phosphates; a nucleoside plus one or more phosphates. dATP, cAMP, and GDP are all nucleotides. The nucleotides in DNA are deoxyadenylate (A), deoxycytidylate (C), deoxyguanylate (G), and thymidylate (T). The nucleotides in RNA are adenylate (A), cytidylate (C), guanylate (G) and uridylate (U). DNA synthesis involves the addition of a nucleotide to the 3´ end of a growing strand; the energy for the reaction comes from the hydrolysis of a nucleotide triphosphate (dATP, dCTP, dGTP, dTTP) to a nucleotide monophosphate (dAMP, dCMP, dGMP, dTMP). There is also a condensation reaction involved in forming the phosphodiester bond.

Oligonucleotide: A short strand of DNA or RNA.

Optical Density: A special term for absorbance when the path length of the light (the thickness of solution through which the light passes) is 1 cm. "Optical density" and "absorbance" are sometimes used interchangeably, because the path length of the light almost always equals 1 cm. (See Spectrophotometry section in Chapter 3.)

Polynucleotide: A strand of DNA or RNA. This term connotes a strand that is shorter than one you would call a nucleic acid.

Residue: A nucleotide that is part of a nucleic acid. (More generically, a residue is what is left of any type of monomer after it has been incorporated into a polymer.) When a nucleotide becomes attached to the 3´ end of a growing strand, the phosphate on the 5´ carbon loses an –OH and the –OH on the 3´ carbon loses its H (this is the condensation reaction that forms the covalent linkage between the nucleotides). The word "residue" describes this H_2O-less mononucleotide.

Specific Activity: Amount of activity of a radionuclide per unit mass of radioactive probe.

ss: Abbreviation for single-stranded DNA or RNA.

DETERMINING THE RELATIVE MOLECULAR WEIGHT OF A DNA OLIGONUCLEOTIDE FROM ITS SEQUENCE

Often in the course of using oligonucleotides, it is necessary to know their molecular weights. If you know the sequence of an oligonucleotide, then all it takes is a simple calculation to determine the molecular weight. The calculation accounts for the molecular weight of each nucleotide residue present, which is just the molecular weight of that residue times the number of times it appears in the oligonucleotide.

Whether your oligonucleotide is single-stranded or double-stranded, you only have to count the number of each residue in one strand, because in a double-stranded oligonucleotide you will know automatically what residues are present in the complementary strand. For example, if you know how many deoxycytidylates (Cs) there are in one strand of a double-stranded DNA sequence, then you automatically know how many deoxyguanylates (Gs) there are in the other strand. For double-stranded DNA (or RNA), you count the nucleotides in one strand, but, instead of multiplying by the molecular weight of each nucleotide to determine the strand's molecular weight, you multiply by the molecular weight of the base pair each nucleotide represents.

When a nucleotide is incorporated into a nucleic acid, it ends up one H_2O smaller than the equivalent isolated nucleotide monophosphate. This happens because the phosphate group attached to the 5´ carbon loses an OH, and the hydroxyl group attached to the 3´ carbon loses an H. The molecular weight of a residue is therefore 18.015 D less than the molecular weight of the nucleotide monophosphate. When you purchase oligonucleotides or primers for a PCR, there is a further correction you have to make because the 5´ nucleotide has no phosphates (it is because of the way oligos are synthesized), and the 3´ nucleotide is not part of a bond; at both the 5´ and 3´ ends, there is an H on each O instead. Finally, at physiological pH, all the phosphates are ionized, so you may wish to take into account that each residue is one hydrogen smaller (although this difference is so small comparatively that it makes no difference compared with other uncertainties, and most calculators and tables ignore it, as does the table below).

Molecular Weights of Nucleotides	M_r
A in DNA = 313.22 D	313.22
C in DNA = 289.18 D	289.18
G in DNA = 329.22 D	329.22
T in DNA = 304.21 D	304.21
A in RNA = 329.22 D	329.22
C in RNA = 305.18 D	305.18
G in RNA = 345.22 D	345.22
U in RNA = 306.2 D	306.2

Molecular Weights of Base Pairs	M_r
A+T = 313.22 + 304.21 = 617.43 D	617.43
C+G = 289.18 + 329.22 = 618.40 D	618.40

Determining the Relative Molecular Weight*
of Single-stranded DNA

$$M_r = (N_C \times 289.18) + (N_A \times 313.22) + (N_T \times 304.21) + (N_G \times 329.22) - 61.96$$

M_r = Relative molecular weight of the single-stranded oligonucleotide
N_C = Number of cytidylates in the single-stranded oligonucleotide

N_A = Number of adenylates in the single-stranded oligonucleotide
N_T = Number of thymidylates in the single-stranded oligonucleotide
N_G = Number of guanylates in the single-stranded oligonucleotide
61.96 = Net difference due to missing 5´ PO_2 and additional 5´ and 3´ Hs

Determining the Relative Molecular Weight* of Single-stranded RNA

$$M_r = (N_C \times 305.18) + (N_A \times 329.22) + (N_U \times 306.2) + (N_G \times 345.22) - 61.96$$

M_r = Relative molecular weight of the single-stranded oligonucleotide
N_C = Number of cytidylates in the single-stranded oligonucleotide
N_A = Number of adenylates in the single-stranded oligonucleotide
N_U = Number of uridylates in the single-stranded oligonucleotide
N_G = Number of guanylates in the single-stranded oligonucleotide
61.96 = Net difference due to missing 5´ PO_2 and additional 5´ and 3´ Hs

Determining the Relative Molecular Weight* of Double-stranded DNA

$$M_r = (N_C \times 618.40) + (N_A \times 617.43) + (N_T \times 617.43) + (N_G \times 618.40) - 123.92$$
or
$$M_r = [(N_C + N_G) \times 618.40] + [(N_A + N_T) \times 617.43] - 123.92$$

M_r = Relative molecular weight of the double-stranded oligonucleotide
N_C = Number of cytidylates in the double-stranded oligonucleotide
N_A = Number of adenylates in the double-stranded oligonucleotide
N_T = Number of thymidylates in the double-stranded oligonucleotide
N_G = Number of guanylates in the double-stranded oligonucleotide
123.92 = Net difference due to missing 5´ PO_2s and additional 5´ and 3´ Hs

DETERMINING DNA CONCENTRATION

Many labs are equipped with instruments that use one method or another to automatically estimate the size of a DNA fragment or quantify the amount of DNA in a sample. If you have such an instrument, of course use it, but be sure to read and follow the instructions carefully. If you do

---

*If you wish to report the molecular weight, rather than the relative molecular weight, just multiply the M_r by 1 g/mole or 1 D (for more on MW vs. M_r, see page 128).

not stay within the limits of the machine's capabilities, you may introduce errors into your experiment. If you don't have an automated system, then the following collection of techniques will describe how to accomplish these calculations by hand. Which technique you decide to use in your own experiments depends on what you need to know about your sample, how precisely you need to measure the sample, and the physical nature of your sample. In all of the techniques described, quantification involves comparing some characteristic of your unknown (its A_{260}, for example) to the same characteristic of a sample of known attributes. In essence, this is what an automated machine does. This section describes how to accomplish the measurements by hand, and how to determine if your solution is contaminated with another molecule that might affect the accuracy of your measurement.

COMPARISON OF METHODS OF NUCLEIC ACID QUANTIFICATION

Method	Property Measured	What It Is Good For	Limitations
Absorbance spectrum	Concentration of all absorbing molecules	Detecting contamination	
A_{260}:A_{280}	Protein contamination in nucleic acid sample	Detecting high levels of protein contamination	Insensitive to low or moderate levels of protein contamination
Spectrophotometry (A_{260})	Concentration	Measurement of pure samples	Insensitive to contamination
Agarose plate method	Nucleic acid concentration	Small contaminants removed from DNA prior to quantification	Insensitive to RNA contamination; generates ethidium bromide waste
Minigel method	DNA concentration	Contaminants, including RNA, removed from DNA prior to quantification	Generates ethidium bromide waste
Hoechst 33258	dsDNA concentration	Measurement of dsDNA concentrations between 10 and 250 ng/ml	DNA must be >1 kb; sensitive to GC content; both photobleaching and quenching can affect measurements
Saran Wrap and ethidium bromide	Nucleic acid concentration	Fast estimate of concentrations as low as 1–5 µg/ml	Sensitive to contaminants
Saran Wrap and SYBR Gold	Nucleic acid concentration	Fast estimates of concentrations as low as 1–5 µg/ml	Expensive; sensitive to contaminants

DETERMINING NUCLEOTIDE AND NUCLEIC ACID CONCENTRATION USING SPECTROPHOTOMETRY

Spectrophotometry can be used to measure the concentration of nucleotides and nucleic acids in solution. Chapter 3 explains the math and how spectrophotometers work. Briefly, spectrophotometers exploit the fact that the amount of light absorbed by a solute is directly proportional to the concentration of solute absorbing the light. So, you put a solution into the cuvette and place the cuvette into the spectrophotometer; the spectrophotometer shines light at the sample and tells you how much of that light was absorbed by the sample. You convert that measurement of absorbance into a measurement of concentration using a standard curve that you make using solutions of known concentrations (for a little more information on standard curves, see Chapter 3).

It is very important to remember, however, that all you are measuring is absorbance; the spectrophotometer cannot tell you *what* is doing the absorbing. It is a good idea to test for contamination before using a spectrophotometer to determine the DNA concentration.

Testing for Contamination (Absorbance Spectrum)

The wavelength of light maximally absorbed by nucleic acids is 260 nm; therefore, to determine the concentration of a nucleic acid solution, you would set the spectrophotometer to read absorbance at 260 nm. However, other things, most notably proteins, also absorb light of the wavelength 260 nm. The spectrophotometer can only tell you how much 260-nm light was absorbed; it cannot identify what components of the sample did the absorbing. Therefore, if the DNA solution is contaminated by protein, you would overestimate the DNA concentration if you only measured A_{260}.

Luckily, proteins and other contaminants have *their* absorbance maxima at different wavelengths. Protein has a maximal absorbance at 280 nm, for example. To determine if there is protein in the nucleic acid, you can measure the absorbance of the sample at 280. If there is an absorbance maximum (i.e., a peak in the graph of absorbance as a function of wavelength) at 280 nm, protein is contaminating the sample. Proteins, however, are not the only absorbing substances that might contaminate a sample of nucleic acids and ruin your estimate of concentration. To determine if substances other than proteins are contaminating a DNA sample, you measure absorbance at a range of wavelengths, instead of just one (A_{260}) or two (A_{260} and A_{280}). If there are maxima/peaks at other wavelengths, it can

mean that the solution is contaminated with other things. Below are listed some common contaminants and their absorbance maxima:

1. If there is significant absorption at $\lambda = 230$ nm (i.e., the absorbance at
   that wavelength is obviously higher than background), the sample is
   probably contaminated with organic compounds, thiocyanates, or
   phenolate ions.

2. If there is significant absorption at $\lambda = 270$ nm, there is probably phenol contamination.

   > **Note:** To characterize protein contamination accurately in a sample, the
   > solution must be free of phenol.

3. If there is significant absorption at $\lambda = 280$ nm, there is probably protein contamination.

4. If there is significant absorption at $\lambda > 330$ nm, there is probably par-
   ticulate matter in the sample.

To get all these numbers in one fell swoop, you do what is called a spectrum: You measure absorbance at many wavelengths (e.g, every 10 nm
from 200 to 360) and graph absorbance as a function of wavelength (or
your spectrophotometer can do it automatically). This graph is the spectrum, and you can easily scan it for evidence of contamination of all kinds.

About A_{260}:A_{280} (OD_{260}:OD_{280})

A common number to calculate is the ratio A_{260}:A_{280} (the ratio of absorbance
at 260 nm to absorbance at 280 nm; this is also written as OD_{260}:OD_{280},
which just means A_{260}:A_{280} with a path length of 1 cm). This ratio is used to
approximate the percentage of absorbing molecules in the DNA sample that
are protein; but it is not a good approximation unless the contamination is
very high. This ratio was actually first used to determine the amount of
nucleic acid contamination in a protein solution. It is very efficient for that
estimate because low levels of nucleic acid contamination have a large effect
on the ratio; the opposite is, unfortunately, not true.

The graph below shows the relationship of A_{260}:A_{280} (y axis) to the percentage of protein in a sample (x axis). A sample of pure nucleic acid will
have an A_{260}:A_{280} of 2. A sample of pure protein will have an A_{260}:A_{280} of
0.57. In between, the relationship is *not linear*. The point is that the solution can be 30% protein and you still could not reliably distinguish the
A_{260}:A_{280} from that of pure nucleic acid. You will think you have a pure

How Much Is Protein ?

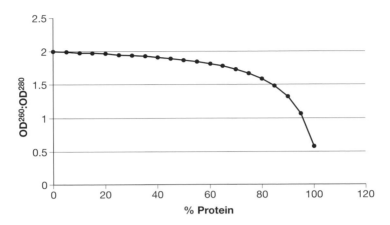

sample, but you won't, and you will overestimate the DNA concentration. So, you really should look at the spectrum, and not just this single number. If the DNA is contaminated, you can process it to remove the protein. If the sample is contaminated but you feel the contamination will not be an issue in further experiments, there are other techniques described later in the chapter that you can use to measure the DNA concentration. If your sample is not contaminated, read on.

Measuring Concentrations of Uncontaminated DNA Solutions Using Spectrophotometry

On average, oligonucleotides and nucleic acids absorb 260-nm light better than they absorb other wavelengths. Since 260 nm is λ_{max} for nucleic acids, 260 nm is the wavelength of incident light that is beamed through solutions of very pure oligonucleotides or nucleic acids to determine the concentration. Individual nucleotides absorb best at other wavelengths so A should be read at λ_{max} for the particular nucleotide if the solution is not a mixture. Note also that pure CTP and pure dCTP should be prepared at pH = 2.0 instead of pH 7.0 (see table below).

For measuring nucleotides and nucleic acids in millimolar concentrations, the relevant version of the Beer-Lambert equation is $A_\lambda = Edc$, where E is the millimolar extinction coefficient. (You are measuring c, the concentration of nucleotides or nucleic acids; thus, your answer is more than likely to be in the millimolar range, not the molar range. You therefore should use the millimolar extinction coefficient.) Rearranging gives $c = A_\lambda / Ed$. Read A_λ (A_λ) off the spectrophotometer; d is the path length (usually 1 cm); then look up *E* (see tables below), which varies with the arrangement of nucleotides.

Note: The nucleotides or nucleic acids must be highly purified, the pH must be carefully kept at 7.0 (unless you are measuring pure CTP or dCTP), the ionic strength of the medium must be low, and the concentration must be greater than 1 mg/liter (1 µg/ml).

Measuring the absorbance (A) of nucleotide samples

To measure the concentration of nucleotides in a very clean solution of a single type of nucleotide, use the following equation.

Note: This technique cannot warn you if the sample is contaminated. Always measure absorbance at a range of wavelengths to test for contamination (see page 133).

$$c = \frac{A_{\lambda\,max}}{E_{NT}\,d} \text{ or, if } d = 1 \text{ cm, } c = \frac{OD_{\lambda\,max}}{E_{NT} \cdot 1 \text{ cm}}$$

c = Concentration [mM]
λ_{max} = Excitation maximum for the nucleotide (see table below)
$A_{\lambda max}$ = Absorbance at the excitation maximum for the nucleotide
$OD_{\lambda max}$ = Optical density at the excitation maximum for the nucleotide
E_{NT} = Millimolar extinction coefficient for the nucleotide [mM^{-1}cm^{-1}] (see table below)
d = Width of the cuvette [cm]. If d = 1 cm, use $OD_{\lambda max}$ instead of $A_{\lambda max}$

Millimolar extinction coefficients

To determine the nucleotide concentration in an uncontaminated solution of only one type of nucleotide, use the following table.

Nucleotide	pH	λ_{max} (nm)	E_{NT} (mM^{-1}cm^{-1})
dATP	7.0	259	15.4
ATP	7.0	259	15.4
DGTP	7.0	253	13.7
GTP	7.0	252	13.7
DCTP	2.0	280	13.1
CTP	2.0	271	12.8
dTTP	7.0	267	9.60
UTP	7.0	262	10.0

☑ Example

Measure a 2.0-ml sample of dATP and get an A_{259} of 0.276. The concentration of dATP in the cuvette is:

$c = 0.276 \div (15.4 \text{ mM}^{-1}\text{cm}^{-1} \times 1 \text{ cm}) = 0.0179 \text{ mM} = 17.9 \text{ μM}$

If the sample in the cuvette is a dilution of another solution, multiply that concentration times the dilution factor to determine the concentration. For example, if the 2-ml sample you measured in the spectrophotometer was a 1:5 dilution of a solution, the stock solution has a concentration of 17.9 μM × 5 = 89.5 μM.

If you need to know how many moles of nucleotide you have, multiply the concentration times the volume of the solution. If you have 2.00 ml of an 89.5 μM solution, you have:

2.00 ml × 89.5 μmoles/liter × 10^{-3} liter/ml = 179 × 10^{-3} μmole = 179 nmoles of nucleotide

If you need to know the concentration in grams per liter (g/liter) or milligrams per milliliter (mg/ml) or micrograms per microliter (μg/μl), use the following conversion:

concentration [μM] × MW of nucleotide × 10^{-6} mole/μmole = concentration [g/liter or mg/ml or μg/μl]

If your concentration is mM, use:

concentration [mM] × MW of nucleotide [g/mole] × 10^{-3} mole/mmole = concentration [g/liter or mg/ml or μg/μl]

Measuring the concentration of uncontaminated oligonucleotides

To calculate the concentration of oligonucleotides in an uncontaminated solution of oligonucleotides based on their absorbance, use the equation below.

Note: This technique cannot warn you if your sample is contaminated. Always measure absorbance at a range of wavelengths to test for contamination (see page 133).

$$c = \frac{A_{\lambda \max}}{E_{oligo}\, d} \text{ or, if } d = 1 \text{ cm, } c = \frac{OD_{\lambda \max}}{E_{oligo} \cdot 1 \text{ cm}}$$

$$
\begin{aligned}
c &= \text{Concentration [mM]} \\
\lambda_{max} &= \text{Excitation maximum for the oligo (see table below)} \\
A_{\lambda max} &= \text{Absorbance at the excitation maximum for the oligo} \\
OD_{\lambda max} &= \text{Optical density at the excitation maximum for the oligo} \\
E_{oligo} &= \text{Millimolar extinction coefficient for the oligo [mM}^{-1}\text{cm}^{-1}\text{] (see table} \\
&\quad \text{below)} \\
d &= \text{Width of the cuvette (cm). If } d = 1 \text{ cm, use } OD_{\lambda max} \text{ instead of } A_{\lambda max} \\
E_{oligo} &= N_A(E_A) + N_G(E_G) + N_C(E_C) + N_U(E_U) + N_T(E_T) \\
N_A(E_A) &= \text{Number of A bases in the oligo x the extinction coefficient for A} \\
&\quad [\text{mM}^{-1}\text{cm}^{-1}] \\
N_G(E_G) &= \text{Number of G bases in the oligo x the extinction coefficient for G} \\
&\quad [\text{mM}^{-1}\text{cm}^{-1}] \\
N_C(E_C) &= \text{Number of C bases in the oligo x the extinction coefficient for C} \\
&\quad [\text{mM}^{-1}\text{cm}^{-1}] \\
N_U(E_U) &= \text{Number of U bases in the oligo x the extinction coefficient for U} \\
&\quad [\text{mM}^{-1}\text{cm}^{-1}] \\
N_T(E_T) &= \text{Number of T bases in the oligo x the extinction coefficient for T} \\
&\quad [\text{mM}^{-1}\text{cm}^{-1}]
\end{aligned}
$$

Millimolar extinction coefficients

To determine the concentration of a solution of oligonucleotides, use the following table:

Base in Oligo	E_{BASE} (λ=260)
A	15.1 mM^{-1}cm^{-1}
G	11.7 mM^{-1}cm^{-1}
C	7.4 mM^{-1}cm^{-1}
T	8.7 mM^{-1}cm^{-1}
U	10.0 mM^{-1}cm^{-1}

☑ Example

Measure a 2.0-ml sample of the oligonucleotide ATT GGC ATC ATC and get an A_{260} of 0.413. The concentration of the solution in the cuvette is:

c = 0.413 ÷ {[(3 × 15.1 mM^{-1}cm^{-1}) + (2 × 11.7 mM^{-1}cm^{-1}) + (3 × 7.4 mM^{-1}cm^{-1}) + (4 × 8.7 mM^{-1}cm^{-1})] × 1 cm} = 0.413 ÷ 125.44 mM^{-1} = 0.00329 mM = 3.29 µM

If the sample in the cuvette is a dilution of another solution, make sure to account for that dilution by multiplying the concentration times the dilution factor. For example, if that 2-ml sample is a 1:10 dilution of a solution, the solution has a concentration of 3.29 µM × 10 = 32.9 µM.

To determine how many moles of oligonucleotide you have, multiply the concentration times the volume of the solution in liters. Continuing the above example, if you have 2.5 ml of a 32.9 μM solution, you have: 2.5 ml x 32.9 μmoles/liter x 10^{-3} liter/ml = 82.25 x 10^{-3} μmoles = 82.25 nmoles of oligonucleotide.

To determine the concentration in g/liter (or mg/ml or μg/μl), multiply concentration [μM] times the molecular weight of the oligo (g/mole) times 10^{-6} mole/μmole.

Measuring the concentration of DNA and RNA

To determine the concentration of DNA or RNA in an uncontaminated solution of nucleic acids using a spectrophotometer, multiply the A_{260} reading by the conversion factors in the following table. These conversion factors are what the concentration (c) would be if A_{260} equaled exactly 1.000; that means they are like unit fractions (see Chapter 1) used to convert A_{260} (which is dimensionless) into a concentration in μg/ml (which is the same as ng/μl). You could say that the units of those conversion factors are μg per ml per "A_{260} reading" or [μg ml^{-1} A_{260}^{-1}].

You may notice that you are not using the Beer-Lambert law and extinction coefficients for this technique, and you are not getting a concentration in molarity. That is because you can't really do that with nucleic acids. The reason is that the absorbing is being done by nucleotides; the spec cannot distinguish whether those nucleotides are part of a polymer or how long a polymer is. In other words, 2 mM dsDNA that was 6000 bp in length would have the same A_{260} reading as 6 mM dsDNA that was 2000 bp in length; two very different concentrations, but the same A_{260}, because the number of absorbing entities is the same.

> **Note:** This technique cannot distinguish between RNA and DNA. Also, it cannot warn you if the sample is contaminated. Always measure absorbance at a range of wavelengths to test for contamination (see page 133).

Converting A_{260} to Nucleic Acid Concentration in μg/ml

To determine the concentration of an uncontaminated solution of nucleic acids, multiply the value of A_{260} by the appropriate conversion factor.

Nucleic Acid	Conversion Factor*
Single-stranded RNA	40 μg/ml
Single-stranded DNA	33 μg/ml
Double-stranded DNA	50 μg/ml

*These conversion factors are appropriate for uncontaminated solutions of nucleic acids over 100 bp in length.

☑ Example

A 2.00-ml sample of ssRNA has an A_{260} = 0.090.

The concentration of ssRNA in the sample is: 0.090 x 40 μg/ml = 3.6 μg/ml

If the sample in the cuvette is a dilution of a solution, multiply the concentration of the sample by the dilution factor. For example, if that 2.00 ml sample was a 1:2 dilution of a solution, the solution has a concentration of 3.6 μg/ml x 2 = 7.2 μg/ml

If you need to know the total amount of ssRNA that you have, multiply that concentration times the volume of the solution. Continuing the above example, if you have 5.00 ml of a 7.2 μg/ml solution, you have:

5.00 ml x 7.2 μg/ml = 36 μg of ssRNA

If you know the MW of your nucleic acid, you can determine the concentration (c) in moles per liter using the following conversion:

c [μg/ml] ÷ MW of nucleic acid [g/mole] x 10^{-6} g/μg x 10^{3} ml/l = c [M]

Approximating DNA Concentrations Using Gel Technology

Quantifying DNA or RNA in a gel is one of those jobs for which there are automated systems. If you do not have such a system, or if you are interested in getting a sense of what it is doing, methods for quantifying by hand (actually by eye) are described here.

Agarose plate method

Note: This technique cannot detect contamination with RNA.

In this method, the fluorescence of ethidium bromide (EtBr) bound to the nucleic acids in the sample you are measuring is compared to the fluorescence from a series of DNA samples of approximately equal length and of known concentrations (the standards). A comparison of the fluorescence intensity of the sample to the fluorescence intensity of the standards allows you to estimate the concentration of the unknown.

In brief, the DNA standards (DNA samples of known concentrations) should be made of DNA from the same species of organism as the DNA being quantified and should be approximately the same length as the unknown. As with any standards, the known concentrations should bracket the expected concentration of the unknown. The standards and the unknown are spotted onto a 1% agarose gel containing 0.5 µg/ml EtBr. A wait of a few hours after the DNA has been spotted gives small contaminants a chance to diffuse away, thus reducing any effect they might have on fluorescence intensity.

METHOD   Outline of Agarose Plate Method

For a more detailed protocol, consult a specialized manual. The basic method is:

1. Spot equal volumes of sample and standards onto the gel.

2. Let it stand at room temperature for a few hours.

3. Photograph using short-wavelength UV.

4. Compare the unknown to the standards.

Minigel method

This method improves on the agarose plate method in that it isolates the DNA from any contaminating RNA. The minigel method works on the same principle as the agarose plate method. Here again, you compare the fluorescence from the unknown sample you are measuring to the fluorescence from a series of known concentrations, the standards. As in the agarose plate method, the standards should be made of DNA from the same species as the unknown and approximately the same size as the unknown. In the minigel method, any contaminating RNA is separated from the DNA sample when the standards and the unknown are run through a gel.

METHOD   Outline of Minigel Method

For a more detailed protocol, consult a specialized manual. The basic method is:

1. Mix 2 µl of DNA with 0.4 µl of loading buffer with bromophenol blue.

2. Run in an EtBr-containing gel until the bromophenol blue has gone 1–2 cm.

3. Destain for 5 minutes in 0.01 M $MgCl_2$.

4. Photograph using short-wavelength UV.

5. Compare the unknown to the standards.

Hoechst 33258 plus Fluorometry

Concentrations of 2–500 ng/ml of dsDNA greater than 1 kb in length can be measured using Hoechst 33258 fluorometry. Hoechst, a German name pronounced "herkst," is a fluorescent dye; the intensity of fluorescence can be quantified using a fluorometer. Unlike spectrophotometry, fluorometry *can* distinguish between dsDNA and other contaminants in the sample. Thus, this is a good technique to use if the sample may be contaminated, or if the sample is mixed with other things on purpose.

Fluorometry can be used to measure DNA concentration even if there are other components in a sample because the Hoechst 33258 dye fluoresces at 458 nm only when bound to dsDNA. More specifically, its fluorescence has different properties when the dye is bound and unbound: Unbound Hoechst 33258 has an excitation maximum at 356 nm and an emission maximum of 492 nm. Under the right conditions, Hoechst 33258 binds in the minor groove of A-T-rich regions of dsDNA greater than 1 kb in length. When it is bound to DNA, its spectral properties change: Bound Hoechst 33258 has an excitation maximum of 365 nm, and an emission maximum of 458 nm. The amount of 458-nm fluorescence is therefore directly proportional to the concentration of dsDNA.

Note: Hoechst 33258 is toxic. Always handle Hoechst 33258 carefully and with gloves, and be sure to follow appropriate safety guidelines. Also, concentrated Hoechst should be kept in a dark container in the refrigerator to maintain its shelf life.

Because the concentration of Hoechst 33258 must be kept within certain limits to obtain an accurate reading, you may have to dilute your sample. If you do, remember to multiply by the dilution factor when you are done. The following conditions need to be met before you start:

1. The fluorometer must be warmed up.

2. The DNA must be bigger than 1 kb.

3. The solution must be free of EtBr.

4. The final concentration of Hoechst 33258 (MW = 533.9) in the sample must be low, 0.5–2.5 μM (or 27–134 μg/ml).

5. This type of fluorometry does not work at extremes of pH, and the emission will be affected by both detergents and salts. The assay must be carried out in 0.2 M NaCl, 10 mM EDTA at pH 7.4.

METHOD Calibrating a Fluorometer

To calibrate the reagents and the fluorometer, you must use known concentrations of high-molecular-weight dsDNA with approximately the same base composition as the unknown (i.e., DNA from the same species). Some fluorometers work like pH meters; i.e., you measure the standards and then adjust the output so that the reported concentration is accurate. Other fluorometers report fluorescence intensity and you must create a standard curve in order to convert intensity to concentration. For example:

1. Measure a range of DNA samples with known concentration (the standards) from 2 to 250 ng of DNA or from 100 to 500 ng of DNA, depending on the expected concentration of the unknown. These standard measurements will give you a graph (a standard curve) telling you the emission at 458 nm as a function of DNA concentration.

2. Measure the concentration of the unknown, or if necessary, measure the fluorescence intensity of the unknown, and then read the DNA concentration off the standard curve. If the emission from the unknown falls outside the range of the standard curve, extend the standard curve by measuring more standards.

Hoechst 33258 can be kept at room temperature as a stock of 0.2 mg/ml in H_2O wrapped in tinfoil. The final concentration of Hoechst 33258 in the sample to be measured should be between 0.5 μM and 2.5 μM (27–134 μg/ml).

1. If you have X unknowns, you will need at least $9X + 18$ ml of fluorometry buffer plus Hoechst 33258. (You need 9 ml per unknown and 18 ml to calibrate the fluorometer; however, it never hurts to make a little extra.) 100 ml is a little more than enough to measure the concentrations of nine unknowns.

2. If you have X unknowns, set up $3X + 6$ glass tubes that can each hold 3 ml.

3. Put 3 ml of fluorometry buffer plus Hoechst 33258 into each tube.

4. Make the standards by adding 100, 200, 300, 400, and 500 ng of reference DNA to five of the tubes. Keep one tube filled with just diluted Hoechst 33258; you will be reading from this tube between *every* reading in order to keep zero at zero (for a procedure cleverly called "zeroing," see Chapter 3, page 63).

5. If the machine automatically converts intensity to concentration, follow the instructions for how to calibrate it. If not, make a standard curve by measuring the fluorescence of each of the standards (do not forget: Between every reading, zero the fluorometer using the pure Hoechst 33258 tube).

6. If you are making a standard curve, graph the reported value of the standards as a function of the concentration of DNA. (Many fluorometers do not report units; in some cases, it is because the value they report is the magnitude of the relative intensity, which is dimensionless.)

7. Zero (by inserting the DNA-less buffer-plus-Hoechst-33258 tube). Add 0.10 µl of the first unknown to a tube with Hoechst, mix by inverting the tube a few times while holding a piece of parafilm tightly to the top of the tube with your thumb, wipe the tube clean with a Kimwipe, insert the tube in the fluorometer, and measure the sample's fluorescence.

8. Zero. Add 1.0 µl of the unknown to a tube with Hoechst, and measure its fluorescence as instructed above.

9. Zero. Add 10 µl of the unknown to a tube with Hoechst, and measure its fluorescence.

10. Read off the amounts of DNA in the unknowns (from the standard curve if necessary). Convert each reading to the same unit (the readings should differ by a factor of 10 because of the dilutions), then average the three readings to estimate your DNA concentration.

The Saran Wrap Method Using Ethidium Bromide or SYBR Gold

The amount of fluorescence emitted by EtBr or SYBR Gold bound to DNA is proportional to the amount of DNA present. You can therefore measure the amount of DNA in a sample by measuring the amount of fluorescence generated when a known amount of dye is combined with a known volume of sample. Like many assays, this one will have you compare the fluorescence from the sample you are measuring to the fluorescence from a series of DNA standards of known concentrations. Observe a match in fluorescent intensity, and you will know the approximate concentration of the unknown. It is critical to realize that this technique can be sensitive to contaminants in the samples. For example, there may be contaminants present in the unknown that will quench or contribute to the fluorescence (see above, Agarose Plate Method). In addition, significant amounts of RNA can make the measurement inaccurate (see above, Minigel Method).

The standards should be made from DNA from the same species of organism as the unknown, and it should be approximately the same size as the size of the unknown. A good set of standards to use to estimate DNA concentration with this method is 0.10, 2.5, 5.0, 10, and 20 µg/ml. The intensities of the fluorescence from standards that are too concentrated will not be distinguishable on film. DNA concentrations that are too low will not be visible at all. The more standards you use, the better you will be able to approximate the concentration of the unknown. Extrapolating a concentration based on fluorescence that falls between two standards of vastly different concentrations is prone to error. Standards keep well for months if stored at −20°C.

METHOD   **Ethidium Bromide or SYBR Gold**

This is a general outline of the method. Consult a more specialized manual for protocols.

1. Spot the standards and the unknowns onto a square of Saran Wrap,

2. Gently mix in an equal volume of stain by pipetting up and down.

3. Photograph the samples using short-wavelength UV for EtBr, or 300 nm light for SYBR gold.

4. Compare the unknown's brightness on the photo to the brightness of the standards, and estimate the DNA concentration.

CALCULATING ENDS PER PICOMOLE OF LINEAR DNA

How many ends you have depends on how many cuts will be made by the nuclease you choose, which, in turn, will depend on the length of the

DNA to be cut and the number of times a cuttable site occurs in that DNA. To calculate the number of ends you have in a sample of cut DNA, use one of the following equations; choose which equation to employ based on whether you know the amount of DNA you have in moles or in grams. Further explanation about where to get some of the numbers follows.

pmole ends = pmoles DNA x 2 x number of cuts

pmole ends = µg DNA x $\dfrac{\text{pmole}}{\text{µg}}$ x 2 x number of cuts

     pmole ends = number of ends (pmole)

     pmole DNA = number of molecules of DNA (pmole)

       µg DNA = weight of the uncut DNA (i.e., concentration ÷ volume) (µg)

   $\dfrac{\text{pmole}}{\text{µg}}$ DNA = species-specific molecular weight of DNA $\left(\dfrac{\text{pmole}}{\text{µg}}\right)$

  number of cuts = predicted number of sites cut by the nuclease

☑ Example

- If you start with 1 pmole of circular DNA molecules (i.e., 6.023×10^{-11} DNA molecules) and do nothing (0 cuts), you will have 1 pmole x 2 x 0 = 0 pmole of ends. Makes sense: If you don't make any cuts, you won't have any ends.

- If you make one cut in each of those molecules, you will have the same number of molecules, each with two ends (every cut creates two ends): 1 pmole x 2 x 1 = 2 pmoles.

- If you cut each molecule in two places, you will have 1 pmole x 2 x 2 = 4 pmoles of ends.

- If you cut in three places, you will have 1 pmole x 2 x 3 = 6 pmoles of ends.

- And so on.

The next dilemma is: How do you know how many cuts you are making in the target DNA? The number of cuts is determined by the endonuclease you choose and the species whose DNA you are digesting. If you assume that the DNA to be digested is randomly arranged, with each base making up exactly 25% of the DNA, you can make a statistical estimate of the number of cuts that will be made based solely on the length of the sequence recognized by the enzyme. The shorter the sequence, the more times it is likely to occur (by chance) in the molecule; and hence, the more cuts the nuclease will make. One way to think of it is if every nucleotide makes up 25% of the DNA, then every position in the DNA has an equal chance of being A T C or G. So, on average, a particular nucleotide will appear every 4 (100% ÷ 25%) nucleotides, a particular pair will appear every 16 (4 × 4) nucleotides, a particular trio, every 64 (4 × 4 × 4), a particular quartet every 256 (4 × 4 × 4 × 4), etc. So, if your endonuclease recognizes a six-nucleotide sequence, you would predict that it would cut every 4096 (4 × 4 × 4 × 4 × 4 × 4) nucleotides. If your DNA is 40,960 bp long, you would predict that this enzyme will make 10 (40,960 ÷ 4096) cuts.

The problem with this reasoning is that different species have different GC contents, i.e., the four bases do *not* each make up 25%, and the percentage they do make up is species-specific. This means that a nuclease that cuts at a sequence with mostly As and Ts will cut a different number of times than a nuclease that cuts at a sequence with mostly Gs and Cs. For example, if the GC content is 60%, that means that G and C each make up 30% of the DNA, and A and T each make up 20%. In this case, you would predict an A every 5 (100% ÷ 20%) nucleotides, a T every 5 nucleotides, a G every 3.33 (100% ÷ 30%) nucleotides, and a C every 3.33 nucleotides. Thus, if your enzyme recognizes ATAT, you would expect it to cut every 625 (5 × 5 × 5 × 5) nucleotides, but if it recognizes CGCG, you would predict it to cut every 123 (3.33 × 3.33 × 3.33 × 3.33) nucleotides, i.e., more frequently.

☑ A Real-life Example

*Bss*HII cuts in the 6-nucleotide sequence GCGCGC and is predicted to make 3×10^5 cuts in the human genome; *Ssp*I cuts in the 6-nucleotide sequence AATATT, but is predicted to make 1.5×10^6 cuts in the human genome. The reason these predictions are so different is that the human genome has more As and Ts than Gs and Cs, so AATATT is likely to occur more often than GCGCGC. The point is, you cannot just go by the length of the sequence that is recognized by the nuclease. To estimate how many cuts you will make, you must take into account the relative amounts of different nucleotides in the DNA you are cutting.

☑ Example

Suppose you want to insert DNA into the plasmid pBR322. This plasmid is 4.361 kb in length, it has a mass of 2.83×10^6 D, and it has one *Bam*HI site in the *TetR* gene (all this information is supplied by the company you buy the plasmid from, or you can find it on their Web site). The tube that you pur-

AVERAGE FRAGMENT SIZE GENERATED

Enzyme	Sequence	Escherichia coli	Mycobacterium tuberculosis	Pyrococcus abyssi
*Apa*I	GGGCCC	68,000	3,000	5,000
*Asc*I	GGCGCGCC	28,000	9,000	350,000
*Avr*II	CCTAGG	290,000	19,000	5,000
*Bam*HI	GGATCC	9,000	3,000	4,000
*Bbv*CI	CCTCAGC	34,000	13,000	11,000
*Bgl*I	GCCNNNNNGGC	2,000	600	6,000
*Bss*HII	GCGCGC	2,000	900	100,000
*Dra*I	TTTAAA	3,000	120,000	2,000
*Eag*I	CGGCCG	16,000	600	11,000
*Eco*RI	GAATTC	7,000	4,000	2,000
*Fse*I	GGCCGGCC	930,000	7,000	140,000
*Hin*dIII	AAGCTT	8,000	18,000	1,000
*Nae*I	GCCGGC	16,000	400	12,000
*Nar*I	GGCGCC	49,000	700	26,000
*Nhe*I	GCTAGC	30,000	6,000	6,000
*Not*I	GCGGCCGC	200,000	4,000	180,000
*Pac*I	TTAATTAA	32,000	No Sites	42,000
*Pme*I	GTTTAAAC	53,000	2,200,000	88,000
*Rsr*II	CGGWCCG	12,000	3,000	140,000
*Sac*I	GAGCTC	31,000	4,000	2,000
*Sac*II	CCGCGG	7,000	700	17,000
*Sal*I	GTCGAC	9,000	1,000	14,000
*Sap*I	GCTCTTC	13,000	22,000	5,000
*Sbf*I	CCTGCAGG	68,000	27,000	290,000
*Sfi*I	GGCCNNNNNGGCC	150,000	4,000	57,000
*Sgr*AI	CRCCGGYG	8,000	900	55,000
*Sma*I	CCCGGG	11,000	2,000	12,000
*Spe*I	ACTAGT	59,000	45,000	8,000
*Sph*I	GCATGC	8,000	3,000	9,000
*Srf*I	GCCCGGGC	120,000	10,000	220,000
*Ssp*I	AATATT	2,000	32,000	2,000
*Swa*I	ATTTAAAT	40,000	4,411,523	29,000
*Xba*I	TCTAGA	120,000	40,000	5,000
*Xho*I	CTCGAG	40,000	3,000	2,000

Reprinted, with permission, from the 2002/03 New England Biolabs Catalog (©NEB).

BY ENDONUCLEASE ACTIVITY

Saccharomyces cerevisiae	Arabidopsis thaliana	Caenorhabditis elegans	Drosophila melanogaster	Mus musculus	Homo sapiens
15,000	42,000	38,000	13,000	5,000	5,000
450,000	920,000	510,000	200,000	280,000	670,000
18,000	20,000	17,000	25,000	4,000	5,000
7,000	8,000	9,000	7,000	4,000	7,000
29,000	29,000	56,000	25,000	7,000	4,000
12,000	26,000	16,000	5,000	5,000	5,000
27,000	120,000	14,000	12,000	19,000	33,000
1,000	900	500	800	1,000	1,000
27,000	37,000	17,000	11,000	30,000	32,000
3,000	3,000	2,000	4,000	4,000	4,000
520,000	610,000	450,000	150,000	160,000	170,000
3,000	2,000	3,000	3,000	4,000	4,000
21,000	25,000	17,000	5,000	22,000	21,000
15,000	47,000	14,000	7,000	17,000	11,000
10,000	12,000	19,000	10,000	7,000	11,000
290,000	610,000	260,000	83,000	120,000	310,000
28,000	8,000	14,000	7,000	34,000	21,000
57,000	37,000	52,000	46,000	100,000	70,000
82,000	77,000	62,000	28,000	120,000	260,000
8,000	5,000	5,000	6,000	3,000	5,000
24,000	24,000	27,000	15,000	23,000	35,000
11,000	12,000	10,000	8,000	48,000	83,000
12,000	12,000	14,000	17,000	9,000	14,000
120,000	120,000	280,000	64,000	23,000	31,000
170,000	460,000	280,000	81,000	23,000	62,000
46,000	34,000	59,000	27,000	98,000	170,000
42,000	40,000	31,000	15,000	7,000	7,000
5,000	5,000	8,000	10,000	9,000	8,000
8,000	9,000	10,000	5,000	4,000	5,000
570,000	No Sites	1,100,000	170,000	120,000	120,000
1,000	1,000	800	900	3,000	1,000
18,000	12,000	8,000	6,000	31,000	15,000
4,000	4,000	4,000	8,000	3,000	4,000
9,000	6,000	9,000	7,000	15,000	22,000

chase contains plasmid at a concentration of 1 mg/ml. You need to know how many ends there will be when you cut 10 μl of plasmid with *Bam*HI.

Because the information you have is the mass of plasmid per volume of solution, use the version of the equation that converts from mass to number, that is:

$$\text{pmole ends} = \text{μg DNA} \times \frac{\text{pmole}}{\text{μg}} \text{ DNA} \times 2 \times \text{number of cuts}$$

1. How many μg of DNA? The volume times the concentration times the appropriate conversion factors (see Chapter 1, page 14 for how to convert units):

   $$10 \text{ μl} \times 1 \text{ mg/ml} \times 10^{-3} \text{ ml/μl} \times 10^3 \text{ μg/mg} = 10 \text{ μg}$$

2. How many pmoles per μg of DNA? One over the molecular weight of the plasmid times the appropriate conversion factors (2.83×10^6 D per molecule means 2.83×10^6 grams per mole):

   $$\frac{1 \text{ mole}}{2.83 \times 10^6 \text{ g}} \times \frac{10^{12} \text{ pmole}}{\text{mole}} \times \frac{10^{-6} \text{ g}}{\text{μg}} = 0.353 \text{ pmole/μg}$$

3. How many cuts? *Bam*HI cuts pBR322 once; so, 1.

4. Plug in the numbers and calculate the answer:

   $$10 \text{ μg DNA} \times 0.353 \frac{\text{pmole}}{\text{μg}} \text{ DNA} \times 2 \times 1 = 7.06 \text{ pmoles of ends}$$

These steps will always work, but if you always use the same vectors and/or inserts, you can make a table and you will never have to calculate again. The following is a table of the picomoles of ends for some commonly used plasmids that have been cut once. To calculate the number of ends for two cuts, multiply the "cut-once" number by 2; three cuts, multiply it by 3, etc.

DNA	pmole ends/μg (if cut once)
Linear pUC18/19	1.14
pBR322	0.706
SV40 DNA	0.58
φX174	0.56
M13mp18/19	0.42
λ phage	0.06
ADD YOUR OWN:	

PREDICTING THE NUMBER OF COPIES OF
TARGET SEQUENCE IN A PCR PRODUCT

The total number of copies of target sequence (including blunt-ended and non-blunt-ended fragments) after any given cycle is the number of template DNA molecules you start with (N_0) times 2 raised to the number of cycles that have occurred (c), i.e., $N_0 \times 2^c$. So after the first cycle, there will be twice as many copies of the target sequence; after the second, there will be four times; after the third, eight times; etc. (see illustration below). Usually, however, the number of blunt-ended fragments is of interest, not the total number of molecules containing the sequence. Here is a picture of the fate of one DNA molecule during the first three cycles of PCR (at each stage, the strands that have just been copied have their 5′ ends indicated):

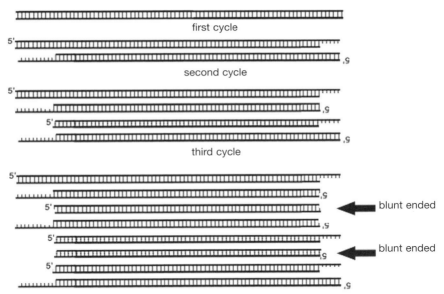

It takes three cycles of PCR (polymerase chain reaction) to create, from a DNA template, blunt-ended fragments that will continue to be amplified throughout the PCR program. That third cycle results in two blunt-ended fragments from each starting molecule; so it is at cycle 3 that PCR product begins to accumulate exponentially. After cycle 3, there will be two blunt-ended fragments (N_{bf}) for every template molecule you began with ($N_{bf} = 2 \times N_0$); after cycle 4, there will be double the number there were after cycle 3 (each is duplicated), plus (2 + 2) new ones made from non-blunt-ended fragments, i.e., there will be (2 × 2) + 4 times the initial number, or $8 \times N_0$. After cycle 5, there will be double the ones made from blunt-ended fragments (2 × 8), plus (2 + 2 + 2) new ones, (2 × 8) + 6 or 22 × N_0. There is a series here: 2, 8, 22, 52, 114, etc. Each of these numbers is 2 raised to the cycle-number-minus-2-times-the-cycle-number. Put symbolically:

$$N_{bf} = (2^c - 2c) \times N_0$$

And as a table:

Cycle	Number of Blunt-Ended Fragments per Template Molecule
3	$2(2^3 - 6)$
4	$8(2^4 - 8)$
5	$22(2^5 - 10)$
6	52
8	240
10	1004 (about 10^3)
15	32,738
20	1,048,536 (about 10^6)
25	33,554,382
30	1,073,741,764 (about 10^9)
35	34,359,738,298
40	1,099,511,627,696 (about 10^{12})

The amount of blunt-ended product continues to accumulate exponentially until the program is complete or one of the reagents becomes limiting, i.e., there is not enough of it to sustain the exponential increase. If a standard PCR is being run with 1 unit of thermostable DNA *Taq* polymerase, the rule of thumb is that the exponential phase lasts until $N_s = 10^{12}$ (N_s = the number of copies of double-stranded target sequence, which after about ten cycles is essentially the same as the number of blunt-ended fragments).

When determining how many blunt-ended fragments have been produced, use the general formula $N_{bf} = (2^c - 2c) \times N_0$ or the above table, but only if the efficiency of PCR is perfect, which of course it is not. So you have to take that into account too. The efficiency is taken into account by a term called Y (see below), which is a value between 0 and 1, and which is determined empirically for each batch of polymerase.

The efficiency of the amplification per cycle is set primarily by the quality of the polymerase, and it can vary significantly. You can calculate the efficiency of the reaction by using one of the DNA quantification methods described above (see page 132) to measure N_0 and N_s and then using the equation below to solve for the value of Y. That value will apply to any subsequent PCRs using that lot of polymerase.

$$Y = \frac{[N_s]^{1/c}}{[N_0]} - 1$$

Y = Efficiency of the amplification per cycle
N_s = Total number of copies of the sequence (measured empirically)
N_0 = Number of copies of the target sequence added at the beginning (measured empirically)
c = Number of cycles

Once you know the efficiency of the polymerase (Y), you can calculate the number of blunt-ended fragments (N_{bf}) after any particular cycle (c) using the following equation:

$$N_{bf} = N_0 (1 + Y)^c$$

N_{bf} = Number of copies of the double-stranded blunt-ended fragment
N_0 = Number of copies of the target sequence added at the beginning
Y = Efficiency of the amplification per cycle for that batch of polymerase
c = Number of cycles

The insert that comes with the polymerase may have information about its efficiency, but it will be reported as the activity, which is some amount of a reference DNA that will be copied in a certain amount of time at a certain temperature.

DETERMINING THE LENGTH OF A DNA FRAGMENT

Electrophoresis

The length (in base pairs or kilobases) of DNA can be determined using agarose or acrylamide gel electrophoresis. The basic principle of DNA electrophoresis is: The longer the piece of DNA, the more slowly it will be pulled through a gel by an electric current. Actually, what you are seeing is the distance the DNA moves, or migrates, through a polyacrylamide or agarose gel in a given amount of time. Bigger nucleic acids do not migrate as far, which is the same as saying they are slower. So, nucleic acids dragged through a gel by a voltage become separated according to size. The reason large DNA fragments move more slowly than small fragments during electrophoresis is that the gel forms a matrix through which the DNA must navigate. Smaller fragments can navigate a gel matrix more easily than large fragments.

The design of your electrophoresis can be adapted to accommodate the approximate size fragments being visualized. For example, to separate DNA fragments that are reasonably small, say anything under 600 bp, you can make a high-percentage agarose gel; say 2% agarose. The high-percentage gel can withstand a higher voltage, which means that you can run the gel at a faster speed and not be concerned that the gel will melt. The higher gel content also enables the distinct separation of smaller, closely sized DNA fragments. To visualize large DNA fragments, say anything over 5 kb, you would run a lower-percentage agarose gel. This loosens the matrix so that small fragments run through very quickly, but larger fragments can also move and separate from each other. You will also lower the voltage at which you run the gel because running large fragments quickly can tangle the long strands and affect separation. If you wanted to separate even larger fragments, say closer to 20 kb, you would run a very-low-percentage agarose gel, perhaps 0.8%, and you would run a very large gel at a very low voltage, say 40 volts, overnight. The long time allowed, combined with a gentle current, will pull large fragments through the gel and separate them. You will, however, lose resolution of small fragments, as they can migrate out of the low-percentage gel into the buffer. Even larger DNA fragments, entire BAC clones or digested MegaYACs, for example, require pulsed-field gel electrophoresis. A pulsed-field gel shifts the angle of the electric field back and forth so as to wiggle the DNA through the gel matrix. For more guidelines on adapting the electrophoresis parameters for ultimate separation of the DNA fragments to translate into accurate fragment size estimations, consult the tables on the facing page and, if necessary, a specialized manual.

At the same time that you run the DNA fragments in a gel, you also run a collection of DNA pieces of known sizes called a DNA ladder. Each "rung" of a DNA ladder is a different length (e.g., each rung is 100 bp longer than the last). You can purchase or make your own DNA ladders, also known as DNA markers, with a difference of 1000 bp between rungs, and ladders that are spaced according to less regular schemes.

When you turn off the voltage at the end running of a gel, you take a picture of the UV-illuminated EtBr-stained gel and compare where your piece of DNA is relative to the rungs of the DNA marker ladder. This gives you an estimate of the size of your DNA. If the DNA fragment migrates to a spot that is between two rungs, you have to extrapolate. Double-stranded DNA molecules migrate at a rate that is inversely proportional to the $\log_{10}$ of the base-pair length of the fragment, meaning that the rungs of the ladder are not evenly spaced. If you estimate the size of the DNA fragment assuming that the relationship between any two *sequential* rungs *is* linear,

Effective range of separation of DNA in polyacrylamide gels

Concentration of Acrylamide (%)	Range Separated (bp)	Size of Fragment Comigrating with	
		Bromophenol Blue (bp)	Xylene Cyanol FF (bp)
3.5	1000–2000	100	460
5.0	80–500	65	260
8.0	60–400	45	160
12.0	40–200	20	70
15.0	25–150	15	60
20.0	6–100	12	45

*Molar ratio of acrylamide:bisacrylamide = 29:1

Range of separation of DNA pieces in different types of agarose

Agarose (%)	Standard	Standard low-EEO*	High Gel Strength	Low Gelling/ Melting Temperature	Low Gelling/ Melting Temperature, Low Viscosity
0.3		5 kb–60 kb			
0.5	700 bp–25 kb				
0.6		1 kb–20 kb			
0.7		800 bp–10 kb			
0.8	500 bp–15 kb		800 bp–10 kb	800 bp–10 kb	
0.9		500 bp–7 kb			
1.0	250 bp–12 kb		400 bp–8 kb	400 bp–8 kb	
1.2	150 bp–6 kb	400 bp–6 kb	300 bp–7 kb	300 bp–7 kb	
1.5	80 bp–4 kb	200 bp–3 kb	200 bp–4 kb	200 bp–4 kb	
2.0		100 bp–2 kb	100 bp–3 kb	100 bp–3 kb	
3.0				500 bp–1 kb	500 bp–1 kb
4.0					100 bp–500 bp
6.0					10 bp–100 bp

*EEO stands for electroendosmosis, which is the migration of uncharged molecules toward the cathode during electrophoresis. EEO interferes with the movement of DNA and RNA: The greater the EEO, the worse the resolution of the bands. The lower the EEO, the better.

Note: A number of the chemicals used in this procedure, including EtBr and acrylamide, are hazardous. Be sure to consult the appropriate literature for safe handling and disposal procedures. In addition, you must use Tris base (not Tris HCl or Trizma) in your gel electrophoresis buffers. Pay attention to expiration dates on prefab gels. Be cognizant of the fact that different forms of DNA migrate differently.

your estimate will be pretty good, depending on how much the fragments are separated. For example, if your band falls exactly halfway between the 600- and 700-bp rungs of a DNA ladder, your DNA fragment is not 650 bp long, but it is likely close (probably about 647 bp). If you are going to estimate this way, always use the two closest rungs of the ladder; if you use rungs that are farther apart (say 500 and 800 for a band that is between 600 and 700), you will be much farther off. This type of estimation is quick and is likely accurate enough to continue with most experiments; however, this type of extrapolation is still only an estimate and, like every measurement, it has a certain amount of error associated with it. If the DNA ladder has rungs that are 1000 bp apart or are separated by irregular intervals, such as those produced by λ*Hin*dIII, the error is likely to increase.

If you need a better evaluation of the size of a DNA fragment, you can measure the distance from each rung of the ladder to some reference point on your gel picture—for example, the bottom of the wells—and then graph the distance of each rung from that point as a function of the size of the rung. Because of the logarithmic relationship of DNA fragment size to distance migrated in a gel, you should graph distance migrated as a function of rung length on a logarithmic scale (most spreadsheet programs have a command that allows you to switch the *y* axis to a log scale): This will give you a straight line rather than a curve. Then, you can measure the distance of the band from the same point, and read its length from your graph, or calculate its length using the equation for the line. Remember that such a graph has to be made for each gel run if you are using this method.

CALCULATING THE AMOUNT OF DNA IN A DNA FRAGMENT RECOVERED FROM A GEL

A common way to isolate a DNA fragment is to cut it out of a larger piece of DNA with a restriction endonuclease, run all the pieces on a gel, and then recover the fragment of interest from the gel. At the end of this procedure, you may need to know how much DNA was recovered in your gel fragment. The actual recovery is going to depend on the cutting efficiency of the restriction enzymes you use, and the efficiency of method you use to extract the DNA from the gel. Because this can vary, it is best to quantify your isolated fragment using one of the methods discussed earlier in this chapter. Often, you will know in advance that you need a certain quantity of your fragment for the next steps in your experiment. To make sure you recover a sufficient amount of fragment (remember that efficiency of recovery will be well below 100%), it is important to know the quantity of fragment-containing DNA that you are loading onto your gel. You will want to start with a significantly larger quantity than the amount of fragment you need, as some will be lost in the cutting and isolation. In this case, you typically know the concentration of the larger piece of DNA that you will digest to get your fragment. Assuming that each molecule of full-length DNA yields one molecule of fragment, you can multiply that concentration times the volume of cut-up DNA, and that will give you the moles of fragment. More commonly, though, you need to recover a particular mass (g) of the fragment to proceed with the next experiment, such as a ligation or a probe labeling. You can use ratios to calculate the amount of full-length DNA you would need to start with to yield a sufficient amount of fragment:

$$\frac{\text{Mass of fragment}}{\text{Mass of full-length DNA}} = \frac{\text{Length of fragment}}{\text{Length of full-length DNA}}$$

This makes sense: The proportion of the mass contributed by the fragment should be about equal to the proportion of length contributed by the fragment. Rearranging gives:

$$\text{Mass of fragment} = \frac{(\text{Mass of full-length DNA})(\text{Length of fragment})}{\text{Length of full-length DNA}}$$

You can also think of this in terms of percentage. If the fragment of DNA contributed 15% to the length of the original fragment, then the amount of that fragment in grams should be approximately equal to 15% of the number of grams originally digested in the reaction. This concept assumes that the base-pair composition across the length of DNA is approximately uniform, which typically is a safe assumption.

Be sure to keep track of your units, and remember that since you are going to lose some of your fragment through the digestion and the isolation process, always start with more than you need.

☑ Example

Assume that you started with 2.00 µg/µl of a 4.20-kb DNA plasmid. You know that you want to isolate a 369-bp fragment, and you know that you will need at least 1 µg of fragment for the ligation in the next step of your experiment. How much of the initial plasmid do you need to start with to get 1 µg of fragment?

1. What is the length of the desired fragment? 369 bp

2. What is the length of the full-length plasmid? 4.20 kb = 4.20×10^3 bp

3. What is the mass of fragment desired? 1.00 μg

4. Plug in the numbers and calculate:

   Using a different rearrangement of the above equation gives:

   X μg of plasmid = 1.00 μg fragment × 4.20 × 10^3-bp plasmid ÷ 369-bp fragment

   X = 1.4 μg of plasmid needed to yield 1.00 μg of fragment.

   Since your plasmid is in solution at a concentration of 2.00 μg/μl, you can calculate how much to use to start with 11.4 μg:

   X μl = 11.4 μg ÷ 2.00 μg/μl

   X = 5.70

   So, you will need to start with 5.70 μl of your plasmid solution to get 1.00 μg of fragment. BUT this assumes 100% efficiency of digestion and isolation, so you will need to start with more than this: 5.70 μl is the *minimum* needed. After isolation, you can see how efficient the process was by measuring the exact concentration by spectrophotometry, or any other method. This will also tell you exactly how much of your isolate you will need to give you the 1.00 μg you need for your ligation:

   X μl = 1.00 μg ÷ actual concentration [μg/μl]

DETERMINING MELTING TEMPERATURE

T_m is the temperature at which 50% of the base pairs in a double-stranded nucleic acid (a helix or a duplex) have come apart into single strands (coils) due to the disruption, by heat, of the hydrogen bonds holding the strands together. T_m is particularly important when you are figuring out conditions for hybridizing with probes and for doing PCR. When hybridizing, you need to know the T_m of the probe:target complex to determine the optimal conditions for your hybridization; when doing PCR, you need to figure out the optimal temperature for annealing the primer to the template.

How much heat it will take to make nucleic acid duplexes fall apart depends on how strongly the strands hold on to each other, which depends on how many hydrogen bonds there are and the strength of those bonds. The number of hydrogen bonds in a sequence of dsDNA depends on the number of A-T base pairs, the number of G-C base pairs, and the number of mismatched base pairs. The strength of the hydrogen bonds depends on the salt and formamide content of the medium containing the DNA.

Because there is a direct relationship between base-pair content and T_m, you can use base-pair content to calculate T_m or you can use T_m to make a reasonable estimate of base-pair content.

Using Base-pair Content to Calculate T$_m$

Different equations are used to calculate T_m. Which equation you use depends on what kind of duplex you are dealing with, on what reagents are in your solutions, and on how accurately you need to know T_m. The following is a compilation of equations that allow you to calculate T_m for different kinds of duplexes. Select the equation that matches the type of duplex you are trying to melt. Alternatively, the Web has many online calculators (see Resources) that will calculate T_m for you; be careful though, because many of them don't tell you how the calculation is done, so you can't be sure that the appropriate equation was used.

Note: The Wallace rule ($T_m = 2°C \times [A+T] + 4°C \times [G+C]$) is often used to get a rough estimate of T_m; be warned, it is a *very* rough estimate.

Double-stranded DNA polynucleotides

$$T_m = 81.5 + 16.6 \log_{10} \left(\frac{[Na^+]}{1.0 + 0.7[Na^+]} \right) + 0.41(\%[G+C]) - \frac{500}{n} - P - F$$

T_m = Melting temperature [°C]
[Na$^+$] = Concentration of sodium ions in the buffer [M]
G = Number of G residues in two strands of the dsDNA
C = Number of C residues in two strands of the dsDNA
%(G+C) = Sum of the number of Gs and Cs in the helix, divided by the total number of nucleotides, i.e., the G+C content
n = Length of the duplex in base pairs
P = 1.00°C × percent mismatched pairs in the duplex [°C]
F = 0.63°C × percent formamide in the solution [°C]

Double-stranded RNA polynucleotides

$$T_m = 78.0 + 16.6 \log_{10} \left(\frac{[Na^+]}{1.0 + 0.7[Na^+]} \right) + 0.7(\%[G+C]) - \frac{500}{n} - P - F$$

T_m = Melting temperature [°C]
[Na$^+$] = Concentration of sodium ions [M]
G = Number of G residues in two strands of the dsDNA
C = Number of C residues in two strands of the dsDNA
%(G+C) = Sum of the number of Gs and Cs in the helix, divided by the total number of nucleotides, i.e., the G+C content
n = Length of the duplex in base pairs
P = 1.00°C × percent mismatched pairs in the duplex [°C]
F = 0.35°C × percent formamide in the solution [°C]

DNA-RNA hybrid polynucleotides

$$T_m = 67.0 + 16.6 \log_{10} \left(\frac{[Na^+]}{1.0 + 0.7[Na^+]} \right) + 0.8(\%[G+C]) - \frac{500}{n} - P - F$$

T_m = Melting temperature [°C]

$[Na^+]$ = Concentration of sodium ions [M]

G = Number of G residues in two strands of the dsDNA

C = Number of C residues in two strands of the dsDNA

%(G+C) = Sum of the number of Gs and Cs in the helix, divided by the total number of nucleotides, i.e., the G+C content

n = Length of the duplex in base pairs

P = 1.00°C x percent mismatched pairs in the duplex [°C]

F = 0.50°C x percent formamide in the solution [°C]

Oligonucleotides

$$T_m = \frac{(298.2 \times \Delta H^\circ)}{\Delta H^\circ - \Delta G^\circ + 593.4 \ln[c]} + 16.6 \log_{10} \left(\frac{[Na^+]}{1.0 + 0.7[Na^+]} \right) - 269.3$$

ΔH° = $\Sigma_{nn}(-8.0$ kcal mole^{-1} x $N_{nn})$ – 8.0 kcal mole^{-1} mismatch^{-1} – 8.0 kcal mole^{-1} end^{-1}

ΔG° = $\Sigma_{nn}(-1.6$ kcal mole^{-1} x $N_{nn})$ + 2.2 kcal mole^{-1} – 1.0 kcal mole^{-1} end^{-1}

N_{nn} = Number of nearest neighbors = number of base pairs – 1 (decrease N_{nn} by 2 for each mismatch or loop)

c = Total molar strand concentration [mole liter^{-1}]

$[Na^+]$ = Concentration of sodium ions [M]

Oligonucleotides less than 100 bases in length

If the cation concentration is <0.5 M, and G+C is 30–70%, the following equation can be used:

$$T_m = 81.5°C + 16.6(\log_{10}[Na^+]) + 0.41(\%[G+C]) - \frac{675}{n} - 1.0 \, m$$

T_m = Melting temperature [°C]

$[Na^+]$ = Concentration of sodium ions [M]

G = Number of G residues in two strands of the dsDNA

C = Number of C residues in two strands of the dsDNA

%(G+C) = Sum of the number of Gs and Cs in the helix, divided by the total number of nucleotides, i.e., the G+C content

n = Length of the duplex in base pairs

m = Percentage of mismatched base pairs

Oligonucleotides 14–70 bases in length

If the cation concentration is less than 0.4 M, the following equation can be used:

$$T_m = 81.5°C + 16.6(\log_{10}[K^+]) + 0.41(\%[G+C]) - \frac{675}{n}$$

n = Number of bases in the oligo
$[K^+]$ = Concentration of potassium ions [M]
$\%(G+C)$ = Sum of the number of Gs and Cs in the helix divided by the total number of nucleotides, i.e., the G+C content

Perfect duplexes 15–20 nucleotides in length

If

- the oligonucleotides are 15 to 20 nucleotides in length,
- there are no mismatches, and
- the solvent is of high ionic strength (e.g., 1 M NaCl or 6x SSC),

you can use the following approximation known as the Wallace rule:

$$T_m = 2°C \times (A+T) + 4°C \times (G+C)$$

where (A+T) is the sum of A and T residues in the oligonucleotide, and (G+C) is the sum of G and C residues in the oligonucleotide.

WARNING: This approximation is not very accurate.

Choosing Hybridization and Annealing Temperatures Based on T_m

Once you have calculated T_m for the reagents, you are ready to select hybridization or annealing temperatures.

Hybridization temperature: 25°C below T_m

PCR Annealing temperature: 3–5°C below T_m

To be really sure, do trial runs from 2°C to 10°C below the T_m of the oligonucleotide with lowest T_m, and see which works best.

For touchdown PCR, start at 3°C above T_m of the best matched oligonucleotide. Touchdown PCR is a technique that improves the specificity of the amplification: The first few cycles are run under more stringent conditions, i.e., at higher temperatures, and then the temperature is reduced one degree per cycle until the temperature "touches down" at T_m; this gives the appropriate sequence a head start.

Using T_m to Calculate Base-pair Composition

To measure the T_m of a duplex directly, measure A_{260} (see page 133) at a variety of different temperatures. Because ssDNA has a higher extinction coefficient than dsDNA, the A_{260} will increase dramatically (30–40% over a very short range of temperatures) when the dsDNA begins to melt. If you graph A_{260} as a function of temperature, you will get a graph that is flat (at the A_{260} for dsDNA), then goes up quickly, then levels off and is flat again (at the A_{260} for ssDNA). Find the midpoint of the rapid increase to determine T_m. From that T_m, calculate the GC content of the duplex.

The following empirically derived equation describes the relationship between T_m and base-pair content if nucleic acids are in a solution that is 0.15 M NaCl and 15 mM sodium citrate and if the G+C content is between 30% and 75%.

Note: The units do not balance in these equations. Think of them as representing the relationship between the magnitudes of the numbers on either side.

$$\%(G+C) = \frac{T_m - 69.3°C}{0.41}$$

G = Number of G residues in two strands of the dsDNA
C = Number of C residues in two strands of the dsDNA
%(G+C) = Sum of the number of Gs and Cs in the helix, divided by the total number of nucleotides, i.e., the G+C content
T_m = Melting temperature [°C]

If you calculate this for all of the possible values of %(G+C), you will find that %(G+C) = T_m – 70. If the test is run under other salt conditions where $[M^+] \le 0.5$ M, use:

$$\%(G+C) = \frac{T_m - 81.5°C - 16.6 \log_{10} [M^+]}{0.41}$$

G = Number of G residues in two strands of the dsDNA
C = Number of C residues in two strands of the dsDNA
%(G+C) = Sum of the number of Gs and Cs in the helix, divided by the total number of nucleotides, i.e., the G+C content
T_m = Melting temperature [°C]
$[M^+]$ = Monovalent cation concentration [M]

SPECIFIC ACTIVITY OF A RADIOLABELED PROBE

If you are using a radiolabeled probe, it is a very good idea to know in advance whether it is radioactively labeled *enough* to give a good signal when you do the hybridization. Knowing this keeps you from waiting

hours or days to find out you cannot visualize the results, and it saves filters (and possibly you) from unnecessary exposure to harsh chemicals
and temperatures associated with hybridizations.

The measurement of the incorporation of radioactivity is called the
specific activity. Specific activity is the amount of radioactivity per unit
mass of probe. To determine the specific activity of a probe, you must
know the radioactivity of the radionuclide you start with, and then you
determine how much of that radioactivity was incorporated into the
probe. You know how much radioactivity you start with because it is listed on the bottle of radioactive isotope. The units are Ci/mole, and from
that you need to calculate how many moles you used by multiplying the
volume used times the concentration, times any conversion factors you
need to make the units match. After you calculate how much radioactivity you added to the labeling reaction, you determine how much (what
percentage) of that radioactivity was actually incorporated into the DNA
probe. That percentage, times the amount of radioactivity you added in
the first place, tells you the specific activity of the labeled probe.

Knowing the maximum specific activity (SA) number for the
radionuclide you are using to label the probe can be a helpful guidepost.
If you calculate the specific activity of the probe and it is higher than the
theoretical maximum, you know you have calculated something wrong.
The maximum possible specific activity of a radionuclide is given by:

$$SA_{max} = 3.132 \times 10^9 / T_{1/2}$$

SA_{max} = Theoretical maximum specific activity (Ci)
$T_{1/2}$ = Half-life of the radionuclide being used (hours) (for a discussion of
$T_{1/2}$, see Chapter 2)

Specific Activity Using Trichloroacetic Acid (TCA) Precipitation

Note: The TCA method only works with probes that are greater than 50
nucleotides long because shorter probes will not precipitate efficiently, and thus
your final calculation will underestimate the specific activity of the probe.

In the following equation, the numerator represents the amount of activity in the probe; the denominator represents the amount of probe. Some
conversion factors are built in to this equation (2.2×10^9 and 1.3×10^3);
their units are given below. These conversion factors let you put numbers
into the equation with the units you are likely to know, i.e., μCi and
μCi/nmole, and have the answer come out in dpm/μg.

$$\text{Specific activity of probe} = \frac{L(2.2 \times 10^9)(PI)}{m + \left[(1.3 \times 10^3)(PI)\left(\frac{L}{S}\right)\right]}$$

Specific activity of probe units = (dpm/μg)

L = Input radioactive label (μCi)

Conversion factor = $2.2 \times 10^9 = 2.2 \times 10^6$ (dpm/μCi) $\times 10^3$ (ng/μg)

PI = Proportion of available label incorporated (cpm in washed filter/cpm in unwashed filter)

m = Mass of DNA template (ng)

Conversion factor = 1.3×10^3 (ng/nmole of dNMPs)

S = Specific activity of the input label (μCi/nmole)

1. Calculate L from information on the bottle of radionuclide (volume used x mCi per volume).

2. PI is the percentage of radioactivity that was incorporated into the probe; you get this number by comparing the amount of radioactivity that is left after a sample has been washed off a filter by TCA to the total radioactivity before washing. The counts coming from the washed sample, divided by the counts coming from the unwashed sample, give you the percentage of the total that was incorporated. The measurement of radioactivity is done in a scintillation counter (see Chapter 3).

3. Calculate the mass of DNA template by multiplying the concentration of template in solution times the exact volume of the samples.

4. Find the specific activity of the input label on the bottle of radionuclide.

Adsorption to DE-81 Filters

This procedure is very similar to TCA precipitation. In this case, you are separating unincorporated probe from incorporated probe via their affinity to a filter, rather than their precipitation in TCA. A key difference in the protocols is that this procedure can be used with probes that are shorter than 50 nucleotides.

$$\text{Specific activity of probe} = \frac{L(2.2 \times 10^9)(PI)}{m + \left[(1.3 \times 10^3)(PI)\left(\frac{L}{S}\right)\right]}$$

Specific activity of probe units = (dpm/μg)

L = Input radioactive label (μCi)

Conversion factor = $2.2 \times 10^9 = 2.2 \times 10^6$ (dpm/μCi) $\times 10^3$ (ng/μg)

PI = Proportion of precursor incorporated (cpm in washed filter/cpm in unwashed filter)

m = Mass of DNA template (ng)

Conversion factor = 1.3×10^3 (ng/nmole of dNMPs)

S = Specific activity of the input label (μCi/nmole)

RESOURCES

Sambrook J. and Russell D.W. 2001. *Molecular cloning: A laboratory manual*, 3rd edition. Cold Spring Harbor Laboratory Press, Cold Spring Harbor, New York. (Has detailed protocols for many methods mentioned here.)

Calculations for Oligonucleotides

http://www.basic.nwu.edu/biotools/oligocalc.html (Will calculate all sorts of useful information about your oligos including T_m, MW, GC content, and more. Definitely worth bookmarking once you've confirmed that you trust it.)

http://alces.med.umn.edu/rawtm.html (Will calculate a T_m for an oligo > 8 nucleotides in length using the method described in Breslauer et al., *Proc. Natl. Acad. Sci.* **83:** 3746–3750 [1986].)

DNA Gels

http://www.dna.caltech.edu/protocols/Elution (Gives protocol for running DNA gels if you want to recover a fragment of a particular size.)

Molecular biology databases: http://www.neb.com/ (Look under "Technical Resource")

Quantifying DNA: http://www.mcrc.com/quantifyingDNA.htm

http://www.invitrogen.co.jp/focus/203084.pdf

Restriction Endonucleases

http://rebase.neb.com/ (Has a complete listing of all known restriction endonucleases, including recognition sequences, methylation sensitivity, commercial availability, and literature references.)

http://tools.neb.com/NEBcutter/index.php3 (Has a program to show where cuts would be made in your DNA sequence by any of the known restriction enzymes.)

Specific Activity of Probe Calculators

http://www.ambion.com/techlib/tips/specific_activity_calculator.html (RNA probes)

http://www.ambion.com/techlib/tips/DNA_specific_activity_calculator.html (DNA probes)

CHAPTER 6

Proteins

TALKING ABOUT PROTEINS

Primary (1°) Structure: The sequence of amino acids in a protein. ...DAIDNCS NELVISHEAD PVWRSAVL... is part of the 1° structure of the pecanex protein (SwissProt accession P18490).

Secondary (2°) Structure: Refers to the α-helix and the β-pleated sheet, two motifs that are commonly found within proteins. These three-dimensional arrangements of stretches of amino acids are stabilized by hydrogen bonds between atoms of the N–C–C "backbone" (as opposed to atoms found in the side chains) and thus can be made from many different 1° structures. When a reference is made to an "α-helical membrane-spanning domain," it is the 2° structure that is being described.

Tertiary (3°) Structure: The three-dimensional conformation of the entire protein (as compared to the 2° structure, which is the conformation of just parts of the protein). The 3° structure is stabilized by interactions among the side chains of the amino acids. When a monomeric protein (a protein composed of a single polypeptide chain) is referred to as "globular," it is the 3° structure that is being described.

Quaternary (4°) Structure: The organization of many polypeptides in a single functional unit. The 4° structure of a protein complex is stabilized by noncovalent interactions between the subunits. The classic example is hemoglobin, which comprises four globin proteins held together into a tetramer by ionic bonds; the organization of the tetramer is the 4° structure.

Amino Acid: A small molecule comprising a central carbon atom that makes its four bonds with a hydrogen, a carboxyl group (–COOH), an amino group (–NH$_2$), and a side chain generically referred to as R. R is usually one of 20 groups (for more information about amino acids and their side chains, see pages 169–170).

Amino Terminus (N-terminus, N-term): The end of a polypeptide or protein that has a complete amino group (see figure under Protein, below).

Carboxyl Terminus (C-terminus, C-term): The end of a polypeptide or protein that has a complete carboxyl group (see figure under Protein, below).

Conformation: An arrangement in space. The three-dimensional geometry of a protein.

Dalton: The unit used to describe the mass of one molecule of a protein. 1 dalton = 1.660539×10^{-27} kilograms; 6.022142×10^{26} daltons = 1 kilogram. Dalton is abbreviated D; kilodalton (10^3 D) is abbreviated kD. One molecule of botulinum toxin has a mass of 150 kD (for more on daltons and molecular weights, see Chapter 2).

Glycosylation: The addition of a sugar group to an amino acid side chain.

Molecular Weight (MW): Also called the molecular molar mass, or the mass of a molecule (units: amu or D). This refers to the mass of the molecular formula, which tells you both the number of atoms of each element and their relative proportions in the molecule. The molecular weight of bovine pancreatic insulin is 5733 D. The term molecular weight also refers to the mass per amount, i.e., g/mole of a substance; however, when talking about proteins, it is used almost exclusively to mean the mass of one molecule; thus the units are daltons.

Peptide Bond: A covalent bond between the C of the carboxyl group of one amino acid and the N of the amino group on another. The type of bond that holds together two amino acid residues within a protein (see figure below).

peptide bond

Polypeptide: A linear polymer of amino acids. Usually used synonymously with protein, although "protein" implies a long polypeptide (see figure below).

Protein: A long polymer of amino acids linked together by peptide bonds (see illustration below).

Neutral residue

Amino terminus

Carboxyl terminus

Residue (Neutral Residue): When an amino acid is incorporated into a polypeptide, it loses one H from its amino group and an –OH from its carboxyl group. What is left is called a residue; it is no longer acidic and therefore is called a neutral residue (see figure above).

DETERMINING THE MOLECULAR WEIGHTS
OF PROTEINS

An important number that is used in descriptions of proteins is the molecular weight (MW). The MW of a protein can be used to help identify the protein in a gel, and it is often used, at least temporarily, in the name of the protein. For example, pp[60src] is a 60-kD protein: one molecule of pp[60src] has a mass of 60 kD; one mole of pp[60src] has a MW of 60,000 g/mole and a relative molecular weight $M_r = 60,000$. The MW of a protein is usually described in daltons (D) or kilodaltons (kD), which is a mass per molecule. The MW of a protein can also give an estimate of the size of the protein, although its conformation will influence volume as well as dimensions. The molecular weight of a protein can be calculated if the sequence is known, or it can be measured.

Calculating the Molecular Weight of a Protein from Its Amino Acid Sequence

If you calculate the number of times a residue appears in the protein, then multiply that number times the molecular mass of the residue, and then add up all those products for all the different amino acids, you will almost have the molecular mass of the protein. Once you have calculated the mass of the entire protein based on the mass of the individual residues, you have to add 18.015 D to account for the extra H on the amino-terminal residue and the extra OH on the carboxy-terminal residue. See prowl.rockefeller.edu/javautilities/pepcalc.htm for an online calculator that will give you the MW of the amino acid sequence you paste or type in, with and without various posttranslational modifications.

Below is the equation if you want to try it yourself:

$$MW_p = 18.015 + \sum_{Ala}^{Val} (n_{nr} \times MW_{nr})$$

MW_p = Molecular weight of the polypeptide [D]

$\sum_{Ala}^{Val}$ = Sum the following calculated for each of the 20 neutral residues

n_{nr} = Number of times that neutral residue appears in the polypeptide

MW_{nr} = Molecular weight of the neutral residue [D]

The following table shows the molecular weights of amino acids and neutral residues:

Amino Acid	Abbreviations 3-letter	Abbreviations 1-letter	MW of Amino Acid (D)	MW of Neutral Residue (D)
Alanine	Ala	A	89.09	71.08
Arginine	Arg	R	174.20	156.19
Asparagine	Asn	N	132.12	114.11
Aspartic acid	Asp	D	133.10	115.09
Cysteine	Cys	C	121.15	103.14
Glutamic acid	Glu	E	147.13	129.12
Glutamine	Gln	Q	146.15	128.14
Glycine	Gly	G	75.07	57.06
Histidine	His	H	155.16	137.15
Isoleucine	Ile	I	131.17	113.16
Leucine	Leu	L	131.17	113.16
Lysine	Lys	K	146.19	128.18
Methionine	Met	M	149.21	131.20
Phenylalanine	Phe	F	165.19	147.18
Proline	Pro	P	115.13	97.12
Serine	Ser	S	105.09	87.08
Threonine	Thr	T	119.12	101.11
Tryptophan	Trp	W	204.23	186.22
Tyrosine	Tyr	Y	181.19	163.18
Valine	Val	V	117.15	99.14

Estimating Kilodaltons Using SDS-PAGE

If all proteins were the same shape and were covered with negative charges, they could be put into an electric field, and the speed at which they migrate toward the positive end of the field would be proportional to the molecular weight of the protein. SDS plus heat and some other reagents make proteins linear and cover them with negative charges. With polyacrylamide gel electrophoresis (PAGE), the proteins are put into the field and the speed at which they move is observed. (Note that a number of the chemicals used in this procedure, including acrylamide, are hazardous. Be sure to consult the appropriate literature for safe handling and disposal procedures.) You must use Tris base (not Tris-HCl or Trizma) in your buffers. Pay attention to expiration dates on prefab gels; proteins will still run in old (bad) gels, but the sizes will be off.

SDS-PAGE shows how far the proteins move, or migrate, through a polyacrylamide gel in a given amount of time. The bigger ones don't migrate as far, which is the same as saying they are slower. So, treated proteins, dragged through a gel by a voltage, are separated out according to size. The unknown proteins run at the same time, in the same gel, as a collection of proteins of known sizes. Then, when the voltage is turned off, the locations of the unknown proteins can be compared relative to the proteins of known size, which are called the protein markers or protein standards. This gives an estimate of how big the proteins are.

Actually, first the unknown proteins (and sometimes the MW standards) must be made visible, for example, by Coomassie Blue staining or western blotting. Then, a picture is taken (by photocopying, scanning, or actually photographing), and, on the picture, the locations of the proteins, relative to the protein standards, are measured. Many labs tape the sizes of the standards onto the refrigerator where the standards are kept. This is an excellent practice; just be sure you are looking at the chart that goes with the set of standards you actually used. To determine the size of your protein, compare its position to the position of the nearest molecular-weight standard. If your protein migrates to a spot that is between two standards, you have to extrapolate. If you need an estimate of the size of your protein that is better than just a ballpark, you'll need to graph the molecular weight of each standard (y axis) as a function of the distance it traveled relative to a reference point in the gel (x axis); a convenient reference point is the bottom of the well. Proteins migrate at a rate that is inversely proportional to the $\log_{10}$ of molecular weight, so your y axis should have a $\log_{10}$ scale. Most spreadsheet programs have a command that allows you to switch the y axis to a log scale by checking a box. Once you have this standard curve, you just measure the distance from that same reference point to your protein band, and read its molecular weight from your graph. Note that this standard graph is only good for that gel; every single gel must include protein standards so that you can create the appropriate standard curve. See www.msu.edu/~venkata1/mwcal.htm for an online calculator that will estimate the size of a protein if you type in the sizes of the two nearest standards and the relative positions of all the bands. The table below shows the linear range of separation of proteins in SDS-polyacrylamide gels.

Acrylamide* Concentration (%)	Linear Range of Separation (kD)
5.0	57–212
7.5	36–94
10.0	20–80
12.0	12–60
15.0	10–43

*Molar ratio of acrylamide:bisacrylamide = 37.5:1

Note: Glycolsylation has a significant impact on protein migration; the MWs of glycosylated proteins will be overestimates of the MW of the unmodified protein.

Determining the Molecular Weight of a Protein Using Gel Filtration Chromatography

Also known as molecular exclusion or gel permeation chromatography, gel filtration chromatography separates proteins based on their size. It is very similar to separating different-sized particles by running all the particles through a sieve of a known size. A sieve separates particles that are smaller than its pore size from particles that are larger than its pore size. Gel beads are like little spherical sieves, all piled up in a column. If a solution is poured over the column, the particles that are too big to get stuck in the sieves will flow through quickly and drip out of the column first. Particles that are small enough to get into the inside of the sieves get hung up for awhile, and so come out of the column later. By knowing the pore size and keeping track of when particles emerge, you can estimate particle size. Actually, the technique can give a finer resolution than just "bigger than pore size" separated from "smaller than pore size." Proteins that are intermediate in size (i.e., not so big as to be completely excluded, but not so small as to be completely inside the beads) will be delayed to a degree that is proportional to their size. So, if you are watching the proteins elute from the column, they will actually emerge in the following order:

- *First*, molecules totally excluded from the beads elute first.

- *Second*, molecules partially inside elute in decreasing order of molecular weight.

- *Third*, molecules smaller than the pore size are entirely inside the beads and elute last.

A lot of vocabulary is associated with this technique. The particle size that is being determined is the molecular weight, also known as the molecular mass (D). The molecular mass of the smallest molecule that is too big to get through the pore is called the gel's exclusion limit. This technique cannot be used to determine the molecular mass of a particle that is larger than the gel's exclusion limit. Particles that are smaller than the exclusion limit will be eluted from the column in order of decreasing size (the smaller the protein, the more likely that some part of it can get hung up in the bead, and therefore the longer it will take for all of it to elute). Thus, the particles elute in decreasing order of size (i.e., biggest first).

The order in which a particle elutes is not characterized by the time, but rather by the amount of solvent it takes to get it out completely; if it takes more solvent, this means that it elutes later. So, although time is not being measured, the volume of solvent that has dripped out is being measured. That volume is called the elution volume, V_e, of the solute:

This is the volume of solute that has dripped out of the column at the point at which a particular solute has dripped out.

To make it possible to compare results from experiments in which different-sized columns are used, the elution volume of the solute is normalized to what is called the void volume (V_0) of the column. The void volume of the column is the volume of fluid it takes to move a too-large protein through the column. The normalized value V_e/V_0 is called the relative elution volume (even though, technically, it is not a volume anymore; it is a proportion). The higher the relative elution volume, the smaller the particle.

METHOD   Determining Molecular Weight Using Gel Filtration Chromatography

1. Measure V_0 by running a protein that is definitely larger than the gel's exclusion limit. Tobacco mosaic virus, or another large-molecular-weight marker that does not interact with the gel, can be used for this purpose. The volume of solvent it takes to get all the protein out the bottom is V_0.

2. Subject proteins of known molecular weights (protein standards) to this technique; then graph the molecular weight (on a log scale, or graph log molecular weight) as a function of relative elution volume. This gives a standard curve (for more on standard curves, see Chapter 3):

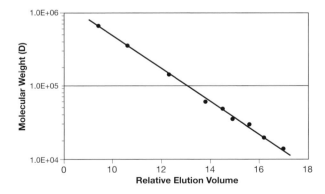

3. To estimate the molecular weight of a protein, determine its relative elution volume, and then read the molecular mass off the standard curve. To calculate relative elution volume, use:

$$V_0 = V_t - V_x$$

V_t = Total volume of the column [liters]
V_x = Volume occupied by gel beads [liters]
V_0 = Void volume (typically 35% of V_t) [liters]

Relative elution volume = $\dfrac{V_e}{V_0}$

Relative elution volume = Column-size-independent measurement of solute behavior

V_e = Volume of solvent required to elute the solute [liters]

The proteins used to create the standard curve must have the same shape as the unknown because globular proteins behave differently from linear proteins.

CONCENTRATION OF PROTEINS

There are a few ways to measure protein concentration. The following are some of the most commonly used assays. Which to use depends on the purity of the protein sample and your equipment.

A_{280} (Spectrophotometry, Ultraviolet Absorbance)

Spectrophotometry allows you to make a rough estimate of protein concentration in the range of 20 µg/ml to 3 mg/ml.

Spectrophotometry can be used to measure the concentration of protein in solution for all the same reasons it can be used to measure the concentration of DNA or any other light-absorbing molecule (see Chapter 3). But it also has the same limitation: It tells you the amount of light absorbed; it cannot tell you the identity of the light-absorbing molecule. So, readings of solutions that are contaminated will be overestimates.

Recall the basic idea of spectrophotometry: The amount of light absorbed by a solution is directly and linearly proportional to the concentration of light-absorbing molecules in that solution. Proteins absorb 280-nm light because tyrosine and tryptophan absorb light of 280 nm; therefore, if you shine 280-nm light into a solution of protein, not all of that light will emerge on the other side. The amount that is missing is proportional to protein concentration (for more information about spectrophotometry, see Chapter 3). Don't forget: Absorbance is very sensitive to the pH and ionic strength of the solution.

METHOD 1 — Using Spectrophotometry to Get a Rough Estimate of Protein Concentration in an Uncontaminated Solution

First, warm up the spectrophotometer for 15 minutes, and then measure the A_{280} (absorbance of 280-nm light) of a group of protein solutions of known concentrations (don't forget about zeroing the spectrophotometer). Next, graph their A_{280} values as a function of concentration; these proteins must have amino acid contents as similar as possible to the protein whose concentration you want to measure. The curve you just made is the standard curve (see Chapter 3). Now, after zeroing the spectrophotometer, measure the absorbance of the unknown, and use the standard curve to convert A_{280} to concentration.

If all you need is a *really* rough estimate, you can forego the pleasure of making a standard curve and use the following very unreliable approximation:

$$0.5 \times A_{280} < c < 2.0 \times A_{280}$$

c = Concentration of protein [mg ml^{-1}]

A_{280} = Absorbance at 280 nm

This technique is only useful for getting a starting estimate of concentration in anticipation of using a more sensitive technique. It is helpful for determining whether the sample to be measured must be diluted before another method is used: If A_{280} is < 2, do not dilute; if A_{280} is > 2, dilution will probably be necessary.

METHOD 2 — Using Spectrophotometry to Get a Rough Estimate of Protein Concentration in a Contaminated Solution

Nucleic acids absorb 280-nm light, just like proteins do. Luckily, 280 nm is the wavelength that is maximally absorbed by proteins, but 260 nm is the wavelength that is maximally absorbed by nucleic acids. So, if you know the solution contains DNA or RNA, you can measure the solution's absorbance at both wavelengths and use the following equation to estimate protein concentration:

$$c = 1.55 \times A_{280} - 0.76 \times A_{260}$$

c = Concentration of protein [mg ml^{-1}]

A_{280} = Absorbance at 280 nm

A_{260} = Absorbance at 260 nm

If you need a more accurate estimate of protein concentration, use one of the following assays (presented in order of increasing accuracy).

The Bradford Assay

The Bradford assay measures 1–200 µg/ml protein concentration.

If you coat proteins evenly with a dye, the amount of bound dye is directly proportional to the amount of protein. If you pick a dye that absorbs light of a particular wavelength (e.g., 595 nm) only when it is bound to protein, then the amount of the 595-nm light absorbed by the solution is a measure of the amount of protein. This is how the Bradford assay works, and the dye in question is Coomassie Brilliant Blue G-250. This assay gets around the problem of contamination, because the only thing that might be in the solution that absorbs 595-nm light is the bound dye. Thus, this assay allows you to quantify protein even if the solution is contaminated.

Why it works: When the Coomassie is free in solution, the wavelength of light it absorbs is 470 nm; however, when the Coomassie is bound to an amino acid, its conformation changes and the wavelength of light it absorbs changes to 595 nm. So, you can simply mix the Coomassie in with the protein, to which it will stick, and then measure A_{595}. Note that some of the chemicals used in this assay are hazardous. Be sure to take the proper safety precautions.

To use A_{595} as a measure of concentration, make a standard curve using a solution of known protein concentration. Many people use solutions of bovine serum albumin, but many have noted problems with it. Catalase, immunoglobulin G, lysozyme, and ovalbumin have all been used successfully; you should use a protein that is as similar as possible to the protein under study in terms of amino acid content. This is because Coomassie Brilliant Blue G-250 binds to arginine, histidine, phenylalanine, tryptophan, and tyrosine; it is the binding of the dye to arginine that is most significant for the assay. So, if your protein is arginine-rich, you should choose an arginine-rich protein to make the standard curve.

Below are two methods for measuring concentrations: a microassay measures 1 µg/ml to 20 µg/ml, and a macroassay measures between 20 µg/ml and 200 µg/ml.

METHOD 1 The Microassay

1. Warm up the spectrophotometer.

2. Make the standard curve: Prepare standards that bracket the range of concentrations you are likely to measure. For the microassay, this

means less than 1.0 µg/ml to greater than 20 µg/ml. For every 1 µl of protein (or protein plus NaOH) solution, add 4 µl of dye solution and let it incubate for 5 minutes. Don't forget to zero the spectrophotometer using a tube of medium plus dye (no protein) between every reading.

3. Zero the spectrophotometer before each reading and measure the A_{595} of the dyed unknowns and determine their concentrations using the standard curve. If the absorbance indicates that the concentration is greater than 20 µg/ml, dilute the sample and measure again. Whatever the dilution factor, multiply the concentration of sample in the cuvette by that number to find the concentration of the original solution.

METHOD 2 The Macroassay

1. Warm up the spectrophotometer.

2. Make the standard curve: Prepare standards that bracket the range of concentrations you are likely to measure. For the macroassay, this means less than 20.0 µg/ml to greater than 200 µg/ml. For every 1 µl of protein (or protein plus NaOH) solution, add 4 µl of dye solution and let it incubate for 5 minutes. Don't forget to zero the spectrophotometer using a tube of medium plus dye (no protein) between every reading.

3. Zero the spectrophotometer before each reading and measure the A_{595} of the dyed unknowns and determine their concentrations using the standard curve. If the absorbance indicates that the concentration is greater than 200 µg/ml, dilute the sample and measure again. Whatever the dilution factor, multiply the concentration of sample in the cuvette by that number to find the concentration of the original solution.

Important Practical Considerations

• Do not use quartz cuvettes; Coomassie sticks to quartz.

• The zeroing of the spectrophotometer is critical for this assay: Use two different buffer blanks to zero the machine; the readings should match.

• Despite what the kit may tell you, the assay does not perform linearly, except over short concentration stretches; the best fit to the standard curve is a second-order equation ($y = ax^2 + bx + c$).

For an online calculator that will analyze your data, see www.msu.edu/ ~venkata1/bradford.htm. The site does not indicate how the data fit the standard curve.

The Lowry Assay (The Hartree-Lowry Reaction)

The Lowry assay measures 2–100 µg/ml protein concentrations.

The Lowry assay is more accurate than the Bradford assay. The original version has been modified, and the new version (the Hartree-Lowry reaction), which is described here, is performed at room temperature, uses fewer reagents, and is the most commonly used version.

This is a spectrophotometric assay that uses the amount of light absorbed by a sample to measure the concentration of the sample. To make the protein absorb light of a particular wavelength, the protein is prepared by a treatment comprising two separate biochemical reactions. The first generates a "Biuret" chromophore, but it is not very sensitive. The second step (the reduction of Folin-Ciocalteu reagent) amplifies the signal about 100x and leads to the creation of a second chromophore, which absorbs 500–750-nm light.

To measure concentration, first create a standard curve showing known concentrations of proteins graphed against their absorbance at 660 nm. (Actually, any wavelength between 500 and 750 nm works fine; use the one least sensitive to contaminants. If the sample may have chlorophyll in it, use 750 nm.) Then, read the concentration of the unknown from the graph. The standard curve is linear in the range of 2–100 µg/ml.

The second step of the preparation is sensitive to the number of tryptophan and tyrosine residues in the protein; so make sure that the standard proteins have a proportion of these amino acids similar to the proteins whose concentrations are being determined. This assay is very sensitive to contaminants that are likely to be in the sample, such as detergents, carbohydrates, TRIS, EDTA, magnesium, and calcium, to name a few.

The Bicinchoninic Acid (BCA) Assay

The BCA assay measures 0.20–50 µg/ml protein concentration.

The BCA assay is more accurate than the Lowry assay because it is not dependent on amino acid composition (at least at high temperatures).

This is a spectrophotometric assay (see Chapter 3) that uses the amount of light absorbed by a sample to measure concentration of the sample. To make the protein absorb light of a particular wavelength, the protein is prepared by a treatment comprising two separate biochemical reactions. The first generates reduced copper (Cu^{+1} from Cu^{+2}). The second step (the interaction with BCA in the presence of protein) leads to the formation of a complex that absorbs 562-nm light.

At low temperatures, the chromophore is generated by the interaction of copper, BCA, and four particular amino acids. At higher temperatures, the chromophore is generated by the interaction of copper, BCA, and peptide bonds. Thus, at higher temperatures (37°C and above), the sensitivity of the assay increases, and the amino acid composition of the protein has less of an effect on the result. To measure the concentration, first create a standard curve showing known concentrations of proteins graphed against their absorbance at 562 nm. Then read the concentration of the unknown from the graph. The standard curve is linear in the range of 1–100 µg/ml. Don't forget to warm up the spectrophotometer and zero between each reading.

Important Practical Considerations

- Water used in the assay must not come through copper pipes.

- Use 1-ml cuvettes for absorbance measurements.

- The samples must be free of reducing agents and copper chelators.

QUANTIFYING PROTEIN ACTIVITY

In addition to wanting to know the amount of protein you have, you might want to quantify what that protein is doing. Specifically, the protein may be part of a chemical reaction, and you might want to describe that reaction. The numbers used to characterize chemical reactions are explained in Chapter 2 and summarized very briefly here:

Consider this generic reaction: $2A + B \leftrightarrow C + 3D$

By convention, A and B are the reactants and C and D are the products, which you know because A and B are on the left and C and D are on the right. The number 2 is the coefficient of A; the number 3 is the coefficient of D. Also by convention, all of the numbers refer to the reaction as written. The characteristics of interest include the equilibrium constant, the reaction rate, the rate constant, and the reaction order.

"I'M A MATHEMATICIAN, AND I'D LIKE SOME
STEROIDS FOR UP HERE."

The Equilibrium Constant

When a reaction is at equilibrium (there is no net change in the concentrations of reactants or products), there is a constant relationship between the concentrations of reactants and products. This relationship between the concentrations of products and reactants in a reaction at equilibrium can be summed up in a single number called K_{eq}. K_{eq} for a given reaction (at a given temperature) is a constant; hence, the equilibrium constant.

K_{eq} is a ratio: The numerator is the product of the concentrations of the products, each raised to a power equal to its coefficient; the denominator is the product of the concentrations of the reactants, each raised to a power equal to its coefficient:

$$K_{eq} = \frac{[C_{eq}][D_{eq}]^3}{[A_{eq}]^2[B_{eq}]} = \frac{\text{(concentration of C at equilibrium)} \times \text{(concentration of D at equilibrium)}^3}{\text{(concentration of A at equilibrium)}^2 \times \text{(concentration of B at equilibrium)}}$$

Because there may be more or fewer than two reactants, an even more generic version is:

$$K_{eq} = \frac{[R_{eq}]}{[L_{eq}]}$$

where $[R_{eq}]$ is the product of the concentrations of products (each raised to its coefficient) on the right-hand side at equilibrium, and $[L_{eq}]$ is the product of the concentrations of reactants (each raised to its coefficient) on the left-hand side at equilibrium.

To determine whether the reaction you are observing at time t is at equilibrium, or going to the right or to the left, calculate:

$$\frac{[R_t]}{[L_t]}$$

where $[R_t]$ is the product of the concentrations of products on the right-hand side at time t, and $[L_t]$ is the product of the concentrations of reactants on the left-hand side at time t; then compare the number to K_{eq}.

If	Then
$\dfrac{[R_t]}{[L_t]} < K_{eq}$	The reaction will go L→R.
$\dfrac{[R_t]}{[L_t]} > K_{eq}$	The reaction will go L←R.
$\dfrac{[R_t]}{[L_t]} = K_{eq}$	The reaction is at equilibrium.

REACTION RATE

The reaction rate is the rate at which the reactants become products. Reaction rates are usually measured as concentration of product appearing per unit time; thus, the units of a reaction rate can be M s^{-1}, but they can also be g s^{-1} or mole s^{-1}, or any other way you would like to measure per second.

To determine a reaction rate, graph time on the x axis and molarity or amount on the y axis, and then take the slope. The slope of the curve describing these data is the reaction rate.

The rate of a reaction can be affected by two things: the concentration of reactants and products and the temperature. The effect of concentrations is summed up numerically as the rate constant and the equilibrium constant.

Concentration and Reaction Rate: The Rate Law, Rate Constants, and the Reaction Order

Take some generic reaction, A + B $\leftrightarrow$ C. The rate of this reaction is the change in the concentration of C over time: $d[C]/dt$. $d[C]/dt$ is a function of the concentration of the reactants [A] and [B]. Below is that function:

$$d[C]/dt = k[A][B]$$

This is the **rate law**, or rate equation, for this one-step reaction (reactions that look different will have different-looking rate equations). k—the proportionality constant that tells the relationship between the reaction rate and the concentration of the reactants—is called the **rate constant**. Another number used to characterize reactions is the **reaction order**.

The order of a reaction is calculated by adding up the exponents on the reactant's concentration terms in the rate equation. For example, for the reaction A + B + 2C $\leftrightarrow$ D, the rate equation is $d[D]/dt = k[A][B][C]^2$, the sum of the exponents is $1 + 1 + 2 = 4$ (recall: if no exponent is written, the exponent equals 1). It is a fourth-order reaction.

Temperature and Reaction Rate

Temperature usually affects the rate of a reaction. To see if a reaction rate varies with temperature, you measure the rate, i.e., measure concentration of product at different times, and then take the slope of the graph of concentration versus time. Then repeat the same measurements at a different temperature, then again at a third temperature, etc. Take the slopes of all of the curves, and then graph the data (*y* axis = slope, *x* axis = temperature). Now, on this new graph of rate versus temperature, you will be able to *see* if there is a relationship between temperature and rate (for more details, see Chapter 2).

V_{max} and K_m: Using Lineweaver-Burke Plots to Characterize Enzyme-catalyzed Reactions That Display Michaelis-Menten Kinetics

Enzyme-catalyzed reactions often do not follow the same patterns as simpler chemical reactions. For example, the rate of the reaction does not always increase with temperature, because either heat or cold can cause an enzyme to denature and therefore stop working. So, characterizing enzyme-catalyzed chemical reactions is a different task from describing simpler chemistry. Have no fear, however; these reactions can be analyzed and characterized relatively simply, at least if they follow a particular pattern known as Michaelis-Menten kinetics, after the two scientists who first described it. The relevant pattern is this: If the relationship of reaction rate (*y* axis) to concentration of substrate (*x* axis) looks like this:

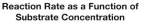

**Reaction Rate as a Function of
Substrate Concentration**

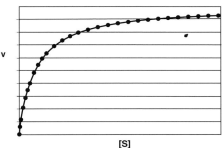

v

[S]

then the enzyme in this graph is following Michaelis-Menten kinetics, and the reaction can be analyzed as described below. This curve is a hyperbola, and its equation is:

$$v = \frac{V_{max}[S]}{K_M + [S]} \quad \text{(the Michaelis-Menten equation)}$$

v = Reaction rate (velocity) [mole s^{-1}]

V_{max} = Maximum theoretical reaction velocity [mole s^{-1}]

[S] = Concentration of substrate [mole liter^{-1}]

K_M = The Michaelis constant; the concentration at which $v = V_{max}/2$ [mole liter^{-1}]

Luckily, the kinetics of many enzyme-catalyzed reactions do look like this because enzymes become saturated. (If the kinetics do not look like this, that's interesting too, because it indicates something else intriguing is happening.) Here is one way to think about saturation: If the test tube (or cuvette) has a certain number of enzyme molecules in it, but no substrate (*x* = 0 on the above graph), the reaction rate will be zero because there is nothing to react. Now, if you add a small bit of substrate, it will be found by enzyme and converted to substrate, so the reaction rate becomes positive. If you add a larger number of substrate molecules, the reaction will proceed faster because enzyme molecules will encounter substrate molecules more often (the higher concentration of substrate molecules means more encounters). If you add an even larger number of substrate molecules, the reaction will proceed faster still. But rate only increases with substrate concentration up to a point, and that point is the point when all the enzymes are, basically, working all the time, i.e., they are saturated.

Once the concentration of substrate is that high, adding more makes a very tiny difference, because the enzymes are already encountering substrate about as frequently as possible, and adding more and more substrate will make less and less difference. That is why the slope of the curve drops precipitously (i.e., the curve flattens). The curve never completely flattens out; what it does is approach an asymptote, namely, the rate at which the reaction would proceed if the substrate concentration were infinite. The value of the reaction rate (if that asymptote could ever actually be reached) would be the absolute fastest the reaction could proceed, i.e., the maximum velocity of the reaction, abbreviated V_{max}. V_{max} (units: mole s^{-1}), is one of the numbers used to characterize an enzyme-catalyzed reaction.

The other number used is K_M, the Michaelis constant. K_M (units: M) is a concentration; specifically, it is the substrate concentration at which the reaction rate is $1/2 V_{max}$. (If you are interested in why this is so, substitute K_M everywhere you see [S] in the Michaelis-Menten equation; that describes the condition where substrate concentration equals the Michaelis constant. Simplify the equation and you will find that when [S] $= K_M$, $v = V_{max}/2$.)

At this point, you might be tempted to graph reaction rate as a function of substrate concentration, then eyeball the value of V_{max}, then divide that in half, and read off the value of substrate concentration that results in a reaction rate of $1/2 V_{max}$ to find K_M. If you do that, your answers could be pretty far off, and, more importantly, there is a very simple transformation of those data that will allow you to make a much better estimate of the values of V_{max} and K_M. This transformation results in what is called a Lineweaver-Burke plot. Like many transformations, this one turns a curve into a line, which is easy to understand. To do this transformation, graph the reciprocal of the reaction rate ($1/v$; units: s mole^{-1}) as a function of the reciprocal of the substrate concentration ($1/[S]$; units: liter mole^{-1}). Applying the transformation to the above data results in a Lineweaver-Burke double-reciprocal plot:

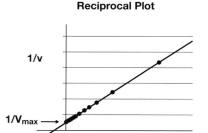

As indicated in the figure above, the y intercept is the reciprocal of V_{max} and the x intercept is the reciprocal of $-K_M$. But don't guess; these numbers can be calculated directly from the equation of the line.

METHOD   Lineweaver-Burke Double-reciprocal Plot

1. Measure the velocity of a chemical reaction given a variety of different substrate concentrations. It should be the initial velocity (the rate as soon as possible after adding the substrate, because as substrate becomes converted, the rate will slow down).

2. Take the reciprocal of each concentration and the reciprocal of each velocity.

3. Graph the reciprocal of velocity as a function of the reciprocal of concentration.

4. Have a statistics program fit a linear regression to the data; the equation of the resulting curve is:

$$\frac{1}{v} = \left(\frac{K_M}{V_{max}} \right) \frac{1}{[S]} + \frac{1}{V_{max}}; \text{ it's a line!}$$

So, the *Y* intercept is $1/V_{max}$ and the slope of the line is K_M/V_{max}.

5. To calculate V_{max}, take the reciprocal of the *y* intercept, and to calculate K_M, divide the slope by the *y* intercept. (K_M is also the negative of the reciprocal of the *x* intercept, but it is easier to find it by dividing slope by *y* intercept.)

Important Practical Considerations

• The enzyme-catalyzed reaction cannot be reversible, i.e., the reverse reaction, where products are converted into reactants, must be negligible.

• The reaction must have, for all practical purposes, a single intermediate.

FLUORESCENCE RESONANCE ENERGY TRANSFER

In addition to absorbing and re-emitting light, fluorescent molecules can absorb light and then excite a second fluorophore via dipole-dipole interactions (i.e., photons are not involved in the second excitation), thus transferring energy via a nonradiative mechanism. For example, if enhanced green fluorescent protein (EGFP) is excited by a photon, it can cause a nearby molecule of Cy3 to fluoresce. This energy transfer is called fluorescence resonance energy transfer or FRET. The energy absorbed by the first fluorophore, the donor, will be transferred to the second fluorophore, the acceptor, if the two fluorophores are within a maximum of about 100 Å (10 nm) of each other. (For examples of useful fluorophore pairs, see the table on page 190.)

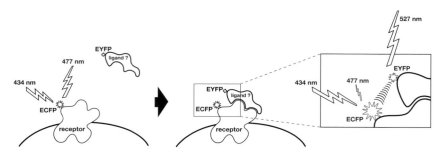

The above figure illustrates, in an oversimplified way, some of the important concepts. In this example, the fluorophore pair is E**C**FP (enhanced cyan fluorescent protein) and E**Y**FP (enhanced yellow fluorescent protein). If the two fluorophores are far apart, exciting the sample with 434-nm light, the excitation maximum for ECFP, causes the ECFP to emit 477-nm light. If, however, the two fluorophores are close together (e.g., the two proteins are interacting), then excitation with 434-nm light leads to 527-nm emission from the EYFP. The ECFP absorbed the 434-nm light, and emitted some 477-nm light, but the amount was greatly reduced because the ECFP transferred some of the energy via resonance to the EYFP. EYFP becomes excited by the transfer of energy, and it emits 527-nm light. If you were watching the ligand approach the receptor using a filter set that allowed you to monitor both 477-nm emissions and 527-nm emissions, you would see the 477-nm light decrease in intensity and the 527-nm light increase in intensity. The closer together the two fluorophores, the less 477-nm light and the more 527-nm light are emitted. FRET thus has two effects: the excitation of the acceptor fluorophore (the EYFP) and the reduction of emission (called "quenching")from the donor fluorophore (the ECFP). Thus, the amount of FRET indicates the relative positions of two fluorophores—the greater the transfer (quantified as the efficiency), the closer the fluorophores.

FRET can be used for a wide variety of applications where data on molecular proximity yield information about a process. Some elegant examples are illustrated at www.probes.com/handbook/boxes/0422.html, and they include the detection of hybridization with molecular beacons, real time PCR, membrane fusion assays, and various reactions in which FRET is used to measure molecular-scale distances.

Measuring Distance Using FRET

The closer the two fluorophores, the more efficient the transfer will be; so, the efficiency of FRET is proportional to the distance between the two fluorophores. Thus, you can use FRET efficiency as a ruler. This ruler can be used to measure the distance between any two fluorophores that are separated by 10–100 Å (1–10 nm). If the donor is conjugated to a protein and the acceptor is conjugated to a different protein, FRET efficiency becomes a ruler that indicates the distance separating the two proteins. If the fluorophores are conjugated to two different parts of the same protein, the ruler indicates the structure of the protein.

FRET is also used to decide whether two proteins interact; the logic is that, for protein interactions to occur, the proteins must get close to each other. For example, the donor fluorophore could be conjugated to a receptor and the acceptor fluorophore could be conjugated to a putative ligand. If FRET occurs, it means the two proteins have interacted, and that is consistent with the hypothesis that they are a receptor-ligand pair. How efficiently the energy is transferred from the donor to the acceptor is called the FRET efficiency, and that is what is measured. A number of factors affect FRET efficiency: the distance between the fluorophores (called r), and a number of other variables that you don't need to be concerned about. These are the refractive index in the range of overlap, the quantum yield of the donor in the absence of the acceptor, the spectral overlap of the two fluorophores, and the orientation of the fluorophores with respect to each other. The reason you don't need to worry about these variables is that the values of most of them are set by the choice of fluorophore pair; so all those values collapse down to a single value that you have to work with, which is called the Forster distance and is given the symbol R_0. R_0 is the distance separating the two fluorophores at which FRET efficiency is 50%. To use FRET as a ruler, measure the efficiency, using one of the techniques described below, and then convert efficiency to a value of distance using the equation described below in Calculating the Distance between Fluorophores.

Important Practical Considerations

- Proteins must be smaller than 100 kD. Proteins larger than 100 kD but smaller than 200 kD can get close enough for FRET to occur, but the likelihood of transfer decreases as size increases; at 200 kD, the likelihood of FRET is low.

- The pH must be constant.

Three different ways to measure FRET efficiency, E, are described below.

Ratiometric Steady-state Fluorescence Intensity Measurement

One way to measure FRET efficiency involves keeping track of a phenomenon called donor quenching. Quenching refers to the fact that if some of the energy emitted by a fluorophore goes to exciting a second fluorophore, the amount emitted as light will decrease (in the figure above, this is the decrease in 477-nm light coming from the ECFP). This decrease in the intensity of light emission is called donor quenching. To determine E this way, measure the fluorescence of each fluorophore under conditions where FRET is not possible (e.g., two samples that have each been labeled with only one of the fluorophores); then measure the fluorescence of each fluorophore under conditions where FRET is happening. The ratio of donor fluorescence to acceptor fluorescence during FRET becomes the numerator of a fraction; the ratio of donor fluorescence to acceptor fluorescence in the absence of FRET becomes the denominator. One minus that fraction equals E:

$$E = 1 - \left(\frac{F_{AD}}{F_D} \right)$$

E = Energy transfer efficiency
F_{AD} = Relative fluorescence yield in the presence of FRET
F_D = Relative fluorescence yield in the absence of FRET

Relative fluorescence yield = ratio of donor fluorescence to acceptor fluorescence

The advantage of using this method is that the comparison of ratios corrects for local variation in fluorescence. The disadvantage of this method is that both donor and acceptor fluorophores may be quenched by other factors, including the process under study, thus causing misleading results. The results can also be difficult to interpret because the ratio is sensitive to the local concentrations of each fluorophore, and those local concentrations may also be affected by the process under study. Thus, it is important to consider the nature and spatial context of the hypothesized interaction.

Donor Quenching and Acceptor Photobleaching

A second method for measuring E is to exploit acceptor photobleaching. This method takes advantage of the fact that fluorescing itself causes fluorophores to lose their ability to fluoresce, a process known as photobleaching. What is being measured is the intensity of donor emission in the presence of acceptor, relative to the intensity of donor emission in the

absence of acceptor. To determine the intensity in the absence of acceptor, photobleach the acceptor by illuminating the specimen with light at
the excitation maximum of the acceptor fluorophore. In practice, you
bleach the acceptor in a small region of the specimen by zooming in and
illuminating with λ_{max} for the acceptor; this can take many minutes. Then
illuminate with λ_{max} for the donor, and measure donor fluorescence
intensity in that region. Finally, zoom out and measure donor fluorescence intensity in a region that has not been bleached. FRET efficiency is
then given by the equation:

$$E_i = 1 - \left(\frac{I_i}{I_{i0}} \right)$$

E = Energy transfer efficiency
I_{i0} = Relative fluorescence yield in the presence of FRET
I_i = Relative fluorescence yield in the absence of FRET

Relative fluorescence yield = ratio of donor fluorescence to acceptor fluorescence

For this technique, it is important to be sure that the photobleached
acceptor does not have any residual absorption at the donor emission
wavelength and that it does not fluoresce at the donor emission wavelength. Also, because photobleaching can take up to 20 minutes, this
technique should be restricted to use on fixed cells, i.e., cells whose components do not move. This technique is useful because it is unaffected by
environmental parameters.

Fluorescence Lifetime Imaging

This third method exploits the fact that in the presence of an acceptor, the
length of time the donor will fluoresce (its lifetime) decreases. To use this
method, just monitor donor fluorescence in the presence and the absence
of acceptor. The lifetime of the donor in the absence of acceptor (determined at the end of the experiment by photobleaching the acceptor) is τ_R;
then τ_i is the lifetime of the donor at pixel i in the presence of acceptor.
The value of efficiency is given by:

$$E = 1 - \left(\frac{\tau_i}{\tau_R} \right)$$

E = Energy transfer efficiency
τ_i = Donor lifetime in the presence of acceptor
τ_R = Donor lifetime in the absence of acceptor

Calculating the Distance between Fluorophores

Once FRET efficiency has been determined by one of the above methods, the distance between the fluorophores is given by:

$$ r = R_0 \times \sqrt[6]{\frac{1}{E} - 1} $$

r = Distance between fluorophores [Å]

R_0 = Forster distance (distance at which FRET efficiency is 50%) [Å]; see table below

E = FRET efficiency

Examples of R_0 for different fluorophore pairs are given in the following table:

Acceptor→ Donor ↓	ECFP	EGFP	EYFP	Cy3	Cy3.5	Cy5
ECFP		47.7 Å	49.9 Å			
EGFP		46.6 Å	55.5 Å	60.0 Å	50. Å	
EYFP			51.1 Å			
Cy3				44.4 Å	51. Å	53. Å
Cy3.5					46. Å	64. Å

An empty box means no available data. 1 Å = 10^{-10} m.

Many other fluorophores can be used. http://www.probes.com/hand book/boxes/0422.html contains links to R_0 tables for the fluorophores they sell, as well as references to literature with more information and more tables.

RESOURCES

Amino Acid Information

http://www.bioscience.org/urllists/aminacid.htm

This site has a table of information about amino acids including abbreviations, molecular weight, pI, CAS registry number, formula, and both line drawing and three-dimensional rendering of the structure of the side chain.

Amino Acid Calculator

A_{280} (spectrophotometry)

http://paris.chem.yale.edu/extinct.html

For any amino acid sequence and A_{280} value, this site will calculate the number and % of each amino acid, the MW, the total number of amino acids, the formal charge, the extinction coefficient (called the molar absorbance), and the concentration, in both mM and mg/ml.

BCA

http://www.ruf.rice.edu/~bioslabs/methods/protein/lowry.html

Bradford Assay

http://www-class.unl.edu/biochem/protein_assay/bradford_assay.htm

Concentration of Proteins

http://www.ruf.rice.edu/~bioslabs/methods/protein/protein.html

FRET

http://www.probes.com/handbook/boxes/0422.html

Hartree-Lowry Assay

http://www-class.unl.edu/biochem/protein_assay/lowry_assay.htm

Molecular Weight of Protein from the Amino Acid Sequence

http://prowl.rockefeller.edu/javautilities/pepcalc.htm

Online calculator in which you paste or type in amino acid sequence, with and without various post translational modifications.

Molecular Weight of Protein from a Gel

http://www.msu.edu/~venkata1/mwcal.htm

Enter the sizes of the two nearest standards and the relative positions of all the bands.

CHAPTER 7

Statistics and Reports: Collecting, Interpreting, and Presenting Numerical Data

INTRODUCTION

The goals of this chapter are to provide:

- A guide to reading other people's commonly used statistics.

- A guide to performing and reporting some of your own statistics.

- A simple explanation of statistics and why they are so useful.

Simply stated, this chapter is about how to plan an experiment, how to describe the results, and how to make and interpret comparisons and measures of relatedness.

Notice the vocabulary is not presented alphabetically as in previous chapters, and the order of the ideas is not typical of statistics books. Rather, both vocabulary and concepts are presented in the order in which you would need to use them over the course of planning and analyzing an experiment.

Caveats

This chapter assumes the following:

- You are using a computer and a statistics program to analyze your data.

- You know how to create a spreadsheet of the data.

- You know how to tell the program to calculate the statistic.

This chapter does not assume that you know what statistic you want to calculate. It does make other assumptions that may not be realistic in your particular situation. Consulting a real statistician, before you begin a project, and then following the advice given, is always the best way to ensure that you will be able to interpret the data when you are finished collecting it. Consulting a statistician after you have the data is the best way of interpreting the data. (Actually, having a statistician to both guide the experiment and then help to interpret the result is probably the very best way.)

As of this writing, statistics packages, like computers, lack brains and free will. They do whatever you tell them to do, even if it's meaningless or misleading. Do not assume that a statistic has meaning just because it can be calculated; do not calculate a statistic just because you can.

WHAT IS STATISTICS?

Statistics, the discipline, is three things: description, probability, and inference. This chapter focuses on more immediately practical statistical information and ignores the fine points of probability.

Inference is based on probability, but you don't have to understand probability as a separate subject to understand how to use inference. All you need to know is that probability theory makes predictions about chance events. You can use those predictions as hypotheses (i.e., the hypothesis that a particular outcome is the result of chance) and then determine whether the data are consistent or inconsistent with that hypothesis. The standard illustration is of dice. Probability predicts that a seven should come up 17% of the time. If the result of an observation is that seven comes up 60% of the time, you would be tempted to reject the hypothesis that chance alone determines which face will appear (i.e., you would conclude that the dice are loaded). However, it is not impossible for a seven to come up 60% of the time by chance; it is just highly unlikely. Statistics tells you how confident to be about rejecting the hypothesis. For most practical purposes, Statistics is:

- A collection of methods for summarizing data so that you can report your results without having to actually list every measurement—descriptive statistics.

- A collection of methods for making reasonable guesses about things you didn't measure based only on things you did—statistical inference.

WHAT ARE STATISTICS?

Statistics, the numbers, are the values you calculate from the data to summarize or compare the data to predictions made from probability theory. Calculating a statistic is, essentially, converting data into a standardized format that makes the data comparable to predictions.

> Explanations about tossing coins and throwing dice illustrate where those predictions come from, and why they are called distributions, and why they are shaped the way they are and have the properties they do. That is the background this chapter largely ignores. The faith you must have to use this chapter is that those predictions *do* make sense, and that calculating a statistic so that you can compare the data to those predictions really does tell you about the data.

DESCRIPTIVE STATISTICS AND STATISTICAL INFERENCE

Descriptive statistics are pictures or calculations that yield numbers that summarize the data. They usually summarize data by indicating where the midpoint is and telling something about the dispersal of values on either side of the midpoint. Common descriptive statistics are the mean and the standard deviation. The mean is a value of the midpoint; the standard deviation describes the spread of values. If the data meet certain criteria, the mean and the standard deviation tell you pretty much all you need to know about the data set. Other descriptive statistics are things like percent similarity and histograms.

Just about everybody uses descriptive statistics: Why report 172 numbers when 2 numbers can contain essentially the same information? Why use a thousand words if one picture will suffice? John Graunt (1662), who has been called the first statistician, put it like this: The point is to "[reduce] great confused volumes into a few perspicuous Tables, and [abridge] such Observations as naturally flowed from them, into a few succinct Paragraphs, without any long Series of multiloquious Deductions... ."

Statistical inferences are calculations that yield numbers that indicate how likely it is that your claims about the data are correct, i.e., they answer those age-old questions: What is the chance that I am right if I claim that the mean I calculated for my specimens represents the actual mean, or what is the chance that I will be wrong if I claim the two treatments yield different results? Statistical inference lets you calculate a number that quantifies how confident you are about the conclusions you've drawn from analysis of your data; reporting these numbers is how you cover yourself. It is a number that means "this is how sure I am, but I could be wrong."

When you use statistical inference to make a comparison, you are making use of the fact that in the past, someone (like Fisher or Dunnet) used probability theory to predict what the collection of results would look like if similar measurements had been made a *very, very* large (let's call it infinite) number of times. These collections of results-from-infinite-theoretical-experiments are called distributions, and the numbers from those distributions are listed in tabular form in the back of statistics books. These are the predictions from probability theory mentioned in the Introduction. Of course, every actual set of measurements has different magnitudes, different units, different numbers of items measured, etc. So, to compare actual data with the distributions tables in the backs of books, you have to put the data into the proper format. The process of making the data comparable to the distributions tables is called calculating a statistic, and that is what many of the equations are about. The different tables are there because the results-from-infinite-theoretical-experiments have different characteristics if the methods used to enumerate the results are different. For example, if the datum is *mean* value of the length of treated individuals, the results-from-infinite-theoretical-experiments will (probably) have what is called a normal distribution; if the data are *proportions* of treated individuals that have gotten longer due to treatment, the results-from-infinite-theoretical-experiments will be in what is called a χ^2 distribution. Again, happily for us, someone else has already calculated all of those distributions and made all of the tables we need, at least for the statistics covered in this book.

Every result within these distributions tables is associated with a number, called the *p*-value, which describes the probability that you would get that result by chance if you were randomly measuring individuals from that distribution. These probabilities, the *p*-values, make intuitive sense. Think of the grades on a first test in a large introductory class. The histogram of grades is the distribution, and, if the class is large enough, it is probably a normal distribution. If you randomly point to a student during a lecture and ask "What did you get on the test?" the student will probably give you a number that is about average. Because most students get about average, you are fairly likely to call on an average student. There is, however, some small probability that you will, by chance, call on the student who got the highest grade. You could look up (a standardized version of) that high number in the distribution and find out exactly what that small probability (the *p*-value) of happening upon that high grade by chance actually is.

Because Fisher et al. have already tabulated the results-from-infinite-theoretical-experiments, you can compare your data to the distributions table, and you can say something about your observations and their relationship to reality. That is pretty great, as it makes possible comparisons that you otherwise could not make.

On a practical level, to use statistics, there are two things you must do before you begin measuring and three things to do after you have finished. Before you start:

1. Determine how many measurements you need to make (see Sample Size and Subject Allocation, p. 208).

2. Decide the best way to make the measurements. This involves deciding in advance how to analyze the data and how to collect the data so that they make that analysis doable (see Choosing Your Statistic, p. 212).

After you have made enough measurements:

1. Describe the data collected (see Descriptive Statistics, p. 202).

2. Do the analysis you planned. Use your computer to manipulate the data to make them comparable to a reference distribution (this is calculating a statistic; see Statistical Inference, p. 220), or use the computer to analyze the relationships among variables (see Relationships, p. 227).

3. Interpret the statistics. Often this means using the computer to calculate the *p*-value, the confidence intervals, or the correlation coefficient.

Because the after-the-measurements calculations are done by the computer, you could say that all you actually have to calculate is the sample size, and all you have to think about is making sure you select the right test and method of data collection. The rest is turning a crank, and nowadays, the computer turns the crank for you. After that, you are done with the statistics, and you are on your own because this is where your knowledge of the system becomes the important analysis tool.

There are many statistical tests to choose from, as you know if you've ever looked in a statistics book. Which test you use depends on certain characteristics of your data, such as the units of your measurement and what you want to know. The information in this chapter is a good place to start, but it is not even close to comprehensive. Subtle things that may not even occur to you to look for, can affect which test is appropriate. Remember, therefore, that consulting a statistician in advance of experimenting is the best plan.

WARNING

If you want to use statistics, it is imperative that your data meet two criteria.

- First, each datum must be independent of all the others. So, the value of the 12th measurement cannot depend on the value of your 11th or 13th or 4th or 40th measurement.

- Second, the data must be from randomly chosen specimens, meaning that every relevant specimen must have an equal chance of being measured. These two characteristics, independence and randomness, are *absolutely essential*, because inference, as mentioned above, is based on probability theory, and probability theory assumes randomness and independence.

  If the data are not random and independent, you will suffer. Maybe not now, maybe not tomorrow, but soon, and for as long as it takes for your competition to stop gloating. If the data do not meet these two criteria (random and independent), STOP HERE. You need statistics that are more sophisticated than those presented in this chapter, and you might be tempted to do something foolish if you continue to read.

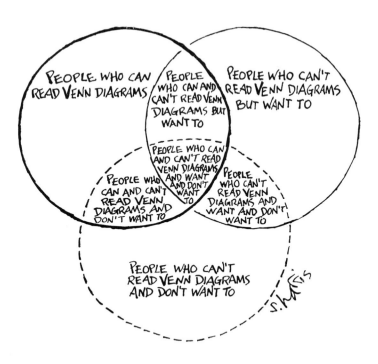

FINAL CHECK BEFORE BEGINNING: CAN THIS CHAPTER HELP YOU?

Probably, but this chapter is not meant to replace either a real statistics book, a statistics course, or a statistician. It is meant to guide you through some of the more common statistics that you might encounter in the literature or wish to use to analyze your own data. It is meant to help you understand the concepts and some simple tests so that you can evaluate the claims others make based on their statistics. But, there are lots and lots of tests, ideas, etc., that are beyond the scope of this book.

Before spending any time searching for an analysis that you will not find in this book, use the following checklist to make sure that your data are the kind this book can help you to describe or to use for inference:

1. The data are random and independent (see the warning above).

2. The data are in a spreadsheet that has one or two columns of numbers.

TALKING ABOUT STATISTICS

Various terms used in this chapter are listed here for quick reference. In addition, some of the terms are defined again in context and others are included here to help you understand the literature, but are not covered in detail. As indicated in the Introduction, the terms are presented in the order you would need to know them to plan and then analyze an experiment. Reading them once through, in order, and then using the list for reference will probably be the most helpful.

Before the Experiment Begins

Statistics: Statistics, the discipline, is three things: probability theory, ways to describe data, and ways to quantify the validity of guesses about all of the whatevers based on measurements of only some of the whatevers.

Data: In this chapter, "data" means "numerical values of measurements"; descriptive numbers. There are different kinds of data:

Discrete, discontinuous—Data that can be reported using whole numbers (integers) only, specifically, counts.

Continuous—Data that require decimal places or fractions to report; measurements that can take any value. Examples: lengths, areas, volumes, ratios, frequencies, and descriptive statistics (means, variances).

Categorical or nominal—Data that can be divided into categories, but when the categories have nothing numerical about them; i.e., there is no particular order to the categories. Example: the effects of different chemicals on some phenotype.

Ordinal—Data that can be divided into categories, where the categories are uneven but can be ordered. Example: young (1–2 months), middle-aged (2–6 months), and old (6 months and older).

Interval—Data that can be divided into evenly spaced ordered categories. Example: length in millimeters. This is the only category for which most statistics are meaningful.

Sample: The subset of whatevers that you measure. The 50 mice you actually weigh. A sample is a collection of replicated measurements, also called trials.

Population: All the whatevers. The 6000 mice that you could weigh and are hoping to learn about by weighing a subset of 50.

n, N, Count: The number of measurements; the sample size.

Statistic: A statistic is a number that quantifies some quality of your sample. Doing statistics usually means performing some calculations with your data to arrive at a particular number that you can then compare to the appropriate reference set. That number is sometimes referred to as a statistic. To compare your data to a standard normal distribution, for example, you would calculate the z statistic. The symbol for a statistic is written in Roman letters, such as $\bar{x}$ and s for mean and standard deviation, respectively.

Parameter: A parameter is a number that quantifies some quality of the entire *population*. A parameter is to the population as a statistic is to your sample. The symbol for a parameter is written in Greek letters, such as μ and σ for mean and standard deviation. Often, you are calculating a statistic to get an estimate of a parameter.

Independent: An *essential* characteristic of good sampling. When sampling is independent, the value of any one datum is in no way affected by the value of any other datum.

Random: An *essential* characteristic of good sampling. When sampling is random, it means that every individual in the population had an equal opportunity of becoming part of the sample that got measured. It is the randomization of sampling that makes most statistics meaningful, because it is randomization that minimizes the effect of unavoidable error, and because probability theory assumes randomness. If your data are not random, nothing in this chapter can help you.

Reference Distribution, Relevant Reference Distribution: The frequency distribution of measurements of nonexperimental individuals to which you will compare your experimental values. A reference distribution is a good thing to compare data to, as long as it is relevant to the data, because it will be subject to the same experimental errors as the data. Frequency distributions of values from controls are relevant reference distributions. When you cannot collect the data for, and then make, your own reference distribution, then you compare your experimental data to the theoretical distributions described below in Distributions.

Paired and Unpaired: These adjectives refer to measurements that come in twos and will be compared, for example, the experimental individuals and the controls. Paired measurements come from individuals that will be compared directly to each other because they are coupled in a meaningful and independent-of-the-experimenter way, e.g., the effect on the operated wing and the effect on the sham-operated contralateral wing. Unpaired means the measures to be compared are not coupled, e.g., 50 sense-strand injected oocytes and 50 antisense-strand injected oocytes. If the data are paired, you will be able to detect differences that you might not be able to detect if the data are unpaired. Differences due to treatment are more likely to show up if the background differences are smaller. Simply put, paired is better.

Blocking: Blocking is self-conscious coupling of unpaired data. It is meaningful coupling, but unlike paired samples, blocked samples are dependent on the experimenter. Planting one control seed and one treated seed in each pot is called blocking. It couples the two subjects in that they will experience the identical environment; therefore, effects of environment are less likely to be confused with effects of the treatment. It *must* still be random, though. For example, the relative positions of the two seeds with respect to each other and the light source must be random from pot to pot. Blocking (with randomization) is good; it is not as good as paired, but it is better than unpaired.

Error, Sampling Error, Experimental Error: This is the unavoidable noise in repeated measurements. You could measure the same thing over and over and still get different values because of how the light is hitting the ruler, or because some days the air conditioning is on and other days it is off, or because you had a cup of coffee at lunch. You cannot ever entirely eliminate error, but if you make sure the error is random, i.e., not likely to push the results in one particular direction, it is usually acceptable. Still, it is important to do everything you can to minimize error, or have a measure of what the error is. A relevant reference distribution is the best way to get a measure of error.

Descriptive Statistics

Frequency Distribution, Histogram: A bar graph with possible values of the measurement on the *x* axis, and the number of times that value occurs in the sample (also known as the data set) on the *y* axis. The *area* of the bar *must be* proportional to the frequency. Do not be tempted to turn the bars in frequency distributions into three-dimensional columns just because it looks more interesting. It is not just distracting, it can actually be deceiving.

Minimum and Maximum: The minimum is the lowest value measured, and the maximum is the highest value measured.

Range: The range is the difference between the smallest and largest values, i.e., the maximum minus the minimum. The range is different from, and bigger than, the spread (see below).

Midpoint (Middle): A characteristic of a data set that is commonly used to describe that data set. Two midpoints are discussed in this chapter: the median and the mean. The median is the value of the point in the middle of the points. The mean is the value in the middle of the values. To get the median, the ordered data are divided into two groups with equal numbers of data points. The median is the value of the point in between the two groups, i.e., half the data points are larger than the median in magnitude and half the data points are smaller than the median in magnitude. To get the mean, the sum of all the values is divided by the sample size.

Spread: A measure of the dispersion of the data around the midpoint. Like the midpoint, the spread is commonly used to describe a data set. Spread is given by IQR, variance, or standard deviation (see below). The IQR is a measure of the spread of the points about the median; the variance and the standard deviation are measures of the spread of values about the mean. Spread and range are not the same thing. Ranges (all the data) are bigger than spreads (a measure of the distribution of the data around the midpoint).

Median: The median is the midpoint of the data in the sense that it divides the data, which have been sorted according to magnitude, into two groups with equal numbers of points. The data in the upper half have values larger than the median; the data in the lower half have values that are smaller. The LD_{50} is a median.

Interquartile Range (IQR): To give a little more information about the spread of the data around the median, the sorted data are divided into subranges, called quantiles, with even numbers of data points (as opposed to even intervals of values) in each quantile. If the data are divided into four quantiles, the quantiles are called quartiles. If the data are divided into 100 quantiles, the quantiles are called percentiles. The IQR is the value at the 3rd quartile minus the value of the 1st quartile.

Outlier: An outlier is a value that is so far away from the other values, it will have undue influence on certain descriptive statistics. If the value of a point is less than $Q_1 - 1.5 \times IQR$ or greater than $Q_3 + 1.5 \times IQR$, that point is an outlier.

Mean: The mean is one kind of midpoint. The mean is the sum of the values divided by the sample size. Most statistics rely on your knowing the mean of your sample; some also require that you know the mean of the population.

Variance: The variance is a measure of the spread of the data around the mean. Its units are the units of the measurement squared. The square root of the variance is the standard deviation.

Standard Deviation: The standard deviation is a measure of the spread of the data around the mean. It is the square root of the variance; thus, its units are the same as the units of the measurement.

Statistical Inference

Inference: Drawing conclusions about the **population** from information about a **sample** taken from that population. Quantifying your uncertainty about your conclusions.

Hypothesis, Hypothesis Testing, Null Hypothesis (H$_0$): You have a hypothesis in mind, unconsciously or consciously, when you decide to do an experiment. It is your guess or prediction about the outcome of your manipulations. The experiment is where you gather the data you will use to test this mental hypothesis, which you will do using statistical inference. So, you can consider statistical inference to be hypothesis testing. Turning your mental hypothesis into a statistical hypothesis requires the same kind of advance planning as choosing how to gather your data; i.e., how to formalize your hypothesis depends on how you want to analyze it. If, for example, you want to use a t-test to quantify your confidence in a conclusion that two means are different, your guess might be that the means are different. But you have a much better chance of *disproving* the hypothesis that they are the *same*; this is the null hypothesis (symbol H_0). (It seems semantic, but it is actually logic.) The way you "statistics-ize" your hypothesis is to state it as the null hypothesis, which is what you do not think the outcome will be. If you believe the means are different, you actually test the null hypothesis that the means (abbreviated $\bar{x}$) are the same. That is written:

H_0 is $\bar{x}_{control} = \bar{x}_{experimental}$

or, if you are quantifying it as the difference between the means

H_0 is $\bar{x}_{control} - \bar{x}_{experimental} = 0$

Reference Distribution: The frequency distribution of measurements of nonexperimental individuals that you collected. A reference distribution is a good thing to compare data to, because it will have the same experimental error as the data. Frequency distributions of values from controls are reference distributions. When you cannot collect the data for, and then make, your own reference distribution, that is when you compare your experimental data to a theoretical reference distribution.

Distributions, Theoretical Reference Distributions: Theoretical reference distributions are frequency distributions that were created based on probability theory rather than on empirical measures. The numbers in the tables in the backs of statistics books come from theoretical distributions with particular shapes and very particular properties; these shapes can each be described by a mathematical function (and are sometimes illustrated in the corner of the table), but in practice they are referred to by name. The distributions were made by using probability theory to predict the outcomes of a very, very large (infinite) number of trials. They are very useful theoretical constructs because they give you something to which you can compare your data if you do not have your own reference distribution, or if you are looking for differences. In other words, theoretical distributions provide you with half a hypothesis.

Particular characteristics of the data type (e.g., raw measurements or proportions or spreads) determine the shape of the theoretical distribution. Here are some basics about some distributions:

Normal (Gaussian) Distribution: A normal distribution is a frequency distribution of values that is shaped like a bell curve. It has its highest point at the mean and is symmetrical about the mean; 95% of the data have values within ±1.96 standard deviations of the mean and 99% of the data have values within 2.58 standard deviations of the mean.

t-Distribution: A *t*-distribution looks like a normal distribution; in fact, it is a variation on a normal distribution. The difference is that you can compare a sample mean to a *t*-distribution even if the population may not be normally distributed and even if you don't know the standard deviation of the population. Means are often compared to *t*-distributions (instead of normal distributions) as a matter of course. When in doubt, comparing your mean to a *t*-distribution is safer, because you have to make fewer assumptions.

Binomial Distribution: A binomial distribution is the frequency distribution for the results of a treatment that can only have two results: succeed or fail, live or die, cause an effect or not cause an effect. It looks like a normal distribution that has had its middle pushed to the right (in technical terms, it is skewed to the right); however, the larger the sample size, the more normal (i.e., less skewed) the binomial distribution becomes.

Poisson Distribution: A Poisson distribution is a special case of the binomial distribution that occurs when the opportunities for the + or – occurrence are ample, but the frequency of + is very low.

χ^2 (Chi Square) Distribution: The χ^2 distribution tells you about the distribution of the variance (as opposed to telling you about the distribution of the mean). A χ^2 distribution looks like a normal distribution that is skewed (pushed to one side).

F Distribution: An F distribution is the distribution of ratios of variances; it looks like a skewed normal distribution.

Distribution Free Tests (Nonparametrics): These are tests where there is no theoretical reference distribution, i.e., no a priori determined distribution of measurement values for a population. Values of measurements for populations are called parameters; hence, nonparametrics or nonparametric statistics. These are useful for analyzing small samples or samples with unknown distributions. Generally speaking, raw numbers are converted to ranks and then analyzed. Some nonparametrics you might encounter are the Wilcoxon test (the statistic you calculate is W) and the Mann-Whitney U test (the statistic you calculate is U), an enhanced version of Wilcoxon's test; both are used when you might otherwise use a *t*-test.

Degrees of Freedom: When calculating a statistic, you are condensing many numbers down to one number. To get that particular value for that particular statistic, you can imagine that *some* of the numbers (values of measurements) were free to have any value. But, once those were set, the rest had to have very particular values, or you would have gotten a different value for the statistic. The number of measurements that are free to have any value represents the degrees of freedom. A good example is the mean, which has $n - 1$ (the sample size minus one) degrees of freedom. Illustration: If the sample size (n) is 4 and the mean is 6, the first three measurements might have been 5, 6, 7; if they were, then the fourth number *must* have been 6 (the average of 5, 6, 7, and 6 is 6). If the first three measurements were 3, 7, 5, the fourth number *must* have been 9 (the average of 3, 7, 5, and 9 is 6). Once the first three are known, the fourth is completely constrained; one number is not free, so $n - 1$ are, hence, $n - 1$ degrees of freedom.

Confidence Intervals: The confidence interval is the region around the sample mean within which the true value of the population mean can be claimed to be found with a particular level of confidence. The region in this case is calculated by multiplying the standard deviation divided by the square root of the sample size times the value of the normal distribution associated with the confidence level of interest: for 95% confidence, $z = 1.96$; for 99% confidence, $z = 2.58$. A confidence interval might be reported as "the mean and 95% confidence interval for population length is 4.26 ± 0.04 mm ($n = 136$)" and it would mean, "we measured 136 individuals and calculated a sample mean of 4.26 mm and a sample standard deviation of 0.24 mm; we wanted 95% confidence in our conclusion about the population; so, the confidence intervals are given by 1.96 x 0.24 mm/$136^{1/2}$ = 1.96 x 0.24/11.7 = 0.040 mm. We therefore believe, with 95% confidence, that the mean length of the population is between 4.22

and 4.30 mm." To make the confidence intervals narrower (i.e., to make the estimate more precise), make the sample size larger.

p-*Value* (or p-*Level*): The probability of a conclusion about a null hypothesis being wrong. Examples: $p = 0.004$, $p \geq 0.99$, $p \leq 0.014$. If $p \leq 0.05$, the level commonly designated as meaning the results are significant, it means there is a 5% chance that you have rejected a hypothesis that is actually correct, i.e., a 95% chance that you were correct to reject the hypothesis. The null hypothesis in question could be "the means of two populations are equal." If $p \leq 0.05$, it means there is only a 5% chance that the two means are equal; so you are justified, according to convention, in claiming that the two means are different.

***Statistically Significant* (p ≤ *0.05*):** This term means that a p-value was calculated to assess the chance that the data are consistent with a null hypothesis, and that p-value was less than or equal to 5%: $p \leq 0.05$. Thus, the null hypothesis was rejected with 95% confidence that it was the right thing to do. This level of uncertainty about the possibility of being wrong is usually considered acceptable (for historical reasons); sometimes $p \leq 0.01$ is preferred. *Important:* If your results are not statistically significant, it does not necessarily mean your mental hypothesis was incorrect and the null hypothesis is supported. $p = 0.10$ or even $p = 0.20$ may be meaningful; it is up to you, with your knowledge of the system, to draw your own conclusions. You may just need a better experimental design to detect a difference that is more subtle than you originally realized. In other words, don't reject your hypothesis until *you*, with all your knowledge of the system, are certain the null hypothesis cannot be rejected. Likewise, you should remember that a $p \leq 0.05$ does not absolutely mean that you can reject the null hypothesis; a 5% chance that the result was due to chance is a 5% chance that the result was due to chance. Again, your judgment matters when you are making the final decision about which hypothesis, yours or the null, to accept.

One- and Two-sided Significance Tests, One- and Two-tailed Distributions: If your **hypothesis** is that two samples are different, but the hypothesis does not include anything about *which one* is bigger, you have to do a two-sided significance test, which is the same as saying you will be comparing your statistic to a two-tailed distribution. You are taking into account that the results could go either of two ways. If your hypothesis does include a prediction about which will be bigger, then you can do a one-sided significance test; the comparison will be to a one-tailed distribution. The p-value for a two-sided test is exactly double the p-value you would get for the same statistic for a one-sided test. Deciding whether your test is one- or two-sided is done before you start, and it requires knowledge of your system.

Analysis of Variance (ANOVA): Everything defined so far (and covered in this chapter) is for analyzing the effect of a single variable. The analyses described allow you to compare two conditions: "in the presence of that variable" versus "in its absence." If you need to make comparisons among three or more states, you need to use analysis of variance.

DESIGNING EXPERIMENTS I: SAMPLE SIZE AND SUBJECT ALLOCATION

If you wish to exploit the power of statistics (and who doesn't?) your data must meet certain criteria. One of those criteria is the correct sample size. Some researchers just try to get as many specimens as possible, or they have some rule of thumb like, always take 11 samples. In fact, the choice of sample size should, if at all possible, be carefully thought out and computed *before* you start your measurements or your experiments. The easiest way to accomplish this is to consult with a statistician *before* you start your measurements or your experiments.

If you are planning to use the mean of your sample size to infer the mean of the population from which that sample was taken, then, the bigger the sample size, the better. This is because, the bigger the sample size, the more precise your estimate of the population mean. (Contrary to some contrarians, you cannot prove anything you want given a big enough sample size; all you can do is get better estimates of the truth.) Unfortunately, there are many reasons why a sample size cannot always be big. It is all right to make the sample size smaller; the tradeoff is that you lose precision. So, to decide how many specimens to measure, you have to decide how much precision you need. The precision in this case is quantified as E, the error. E is the amount by which you are willing to be off. In other words, if you are willing to accept an estimate that is within 6 of the mean, i.e., the mean ± 6, then $E = 6$. You also have to decide how confident you want to be that the population mean is, in fact, within those boundaries. Most of the time scientists want to be either 95% confident or 99% confident. Luckily, there is a formula. For the sample size for result ± E:

$$n = \left[\frac{1.96 \times s}{E} \right]^2 \quad \text{if you want to be 95\% certain that the mean is within } E \text{ of the estimate}$$

$$n = \left[\frac{2.58 \times s}{E} \right]^2 \quad \text{if you want to be 99\% certain that the mean is within } E \text{ of the estimate}$$

Always round up (even if you only theoretically need 0.4 more samples, you can only actually measure an entire sample).

n = Sample size
s = Standard deviation of the sample
E = Amount by which you are willing to be off
1.96 and 2.58 = z scores for 95% and 99% confidence, respectively

To use this equation, you need to know the standard deviation of the sample, which means you need to know the outcome of the experiment before you have performed it. One way around this is to make an educated guess about the standard deviation of the sample. For example, if you have done similar measurements before, you may have a good idea what the standard deviation of the sample is going to be. After you are done, you can check back to make sure you measured enough samples.

The better way is to do a pilot study. In theory, the sample size of the pilot study does not matter, as long as it is greater than 3; however, the bigger the pilot study, the better the first estimate of the standard deviation will be, and the more likely you will be to actually find the minimum necessary sample size. Once you have the data from the pilot study, input them into your statistics program and have the program tell you the standard deviation. Most statistics programs expect the sample to be input as a column of numbers. If you can input your data as a single column of numbers, it is "univariate." Now use the above equation and a calculator to determine the size of the sample you will need to be able to say, with either 95% or 99% confidence, that the mean of the population equals the mean of the sample, plus or minus E.

☑ Example

Assume you have done a pilot study and your statistics program calculates that the mean of the sample is 1254 and the standard deviation is 98. If you want to be 95% sure that the true mean will be within 90 of the sample mean (the mean of the population = 1254 ± 90), then the sample size can be as small as 5 because:

$$n = \left[1.96\,\frac{s}{E} \right]^2 = \left[1.96 \times \frac{98}{90} \right]^2 = 4.6; \text{ round up to } 5$$

If you want to be 99% sure that the true mean will be within 90 of the sample mean, then the sample size must be 8:

$$n = \left[2.58\,\frac{s}{E}\right]^2 = \left[2.58 \times \frac{98}{90}\right]^2 = 7.9;\ \text{round up to } 8$$

If you want to be 99% sure that the true mean will be within 20 of the sample mean (the mean of the population = 1254 ± 20), then your sample size must be 160:

$$n = \left[2.58\,\frac{s}{E}\right]^2 = \left[2.58 \times \frac{98}{20}\right]^2 = 159.8;\ \text{round up to } 160$$

That is quite a jump; so at this point, you might decide that you are willing to be only 95% sure that the true mean will be within 20 of the sample mean ($E = 20$). If you are willing to give up that reliability, then the sample size will drop to 93:

$$n = \left[1.96\,\frac{s}{E}\right]^2 = \left[1.96 \times \frac{98}{20}\right]^2 = 92.2;\ \text{round up to } 93$$

You can see that it is a process of balancing your need for precision (E), your need for confidence (95% vs. 99%), and your ability to collect data (n).

Allocating Subjects to Treatments

In a common experimental scenario, you have a certain number of samples or subjects on which you can experiment and you have to choose how to allocate them among the different treatments. You always have at least two treatments—the experimental and the control. You should probably have at least three—the experimental, the positive control, and the negative control. For certain experiments, it is advisable to have a third control—the no treatment control—meaning four different treatments.

To allocate a fixed number of subjects among different treatments, whose means you intend to compare, the ratio of the numbers put into each treatment should be the same as the ratios of the variances of those treatments. Variance is another measure of the spread of data around the mean; it is the standard deviation squared.

$$n_A : n_B : n_C = s_A^2 : s_B^2 : s_C^2$$

n_J = Sample size for treatment J
$s^2{}_J$ = Variance determined from pilot study

Usually, the values of the experimental data vary more than those of the control data. Another way to think of this is that with equal sample sizes, the control values are going to be more tightly clustered around the mean than will the experimental values. This means that your estimate of the control value for the whole population is more precise, i.e., has a smaller error than your estimate of the experimental value for the whole population. That is a problem because to compare the mean of the experimental values to the mean of the control values, the standard deviations (which are proportional to the uncertainty) must be approximately equal. So, the trick is to figure out how to adjust the sample sizes so that the standard deviations will be the same. This is pretty straightforward. The ratio of the sample sizes (n) in each treatment should be the same as the ratios of the variances (s^2) of those treatments:

$$n_A : n_B : n_C = s_A^2 : s_B^2 : s_C^2$$

So, once again, you have to do a pilot study to know where to begin.

METHOD Allocation of Subjects

1. Do a pilot study

2. Have your statistics program calculate the variance for each treatment.

3. Calculate the ratio of the variances (divide each variance by the value of the lowest variance).

4. Allocate the remaining subjects in the same ratio; divide the number of remaining subjects by the sum of the ratios, and then multiply that quotient times each ratio.

☑ Example

Assume that you have 41 subjects and that you use 16 of them in the pilot study: 8 experimental subjects and 8 controls. The pilot study, in which there were two treatments, yielded the following variances:

Step 2: $s_{exp}^2 = 93$ and $s_{con}^2 = 61$

Step 3: $\dfrac{93}{61} : \dfrac{61}{61} = 1.5:1$

Step 4: 25 remaining subjects.
$25/(1.5 + 1) = 10$

$1.5 \times 10 = 15$ experimental subjects
$1 \times 10 = 10$ control subjects

"ALTHOUGH IT DIDN'T HELP ANY ACTUAL PATIENTS, OUR STATISTICIANS HAVE BEEN ABLE TO PROVE IT'S A VERY EFFECTIVE MEDICINE."

At this point, you may want to do a quick calculation to see if your results will be precise enough for you to draw a conclusion that is meaningful. To determine the error associated with the mean, use:

$$E = \frac{1.96 \times \sqrt{s_{exp}^2}}{\sqrt{n_{exp}}}$$

For this example:

$$E = \frac{1.96 \times \sqrt{s^2}}{\sqrt{n}} = \frac{1.96 \times \sqrt{93}}{\sqrt{15}} = 4.88$$

If you are satisfied with the result that you can be 95% confident that the actual mean of the population is within 4.9 of the sample mean, then go ahead and do the experiment.

DESIGNING EXPERIMENTS II: CHOOSING YOUR STATISTIC

The following is an illustration of where statistical inference fits into experimentation and where probability dictates procedure:

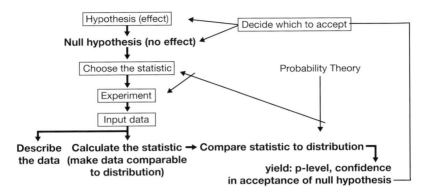

This figure above shows that probability theory is not part of the loop, but it is where distributions come from and therefore it dictates how to analyze the data and therefore how to gather the data. The loop starting at "Hypothesis (effect)" and leading to "Decide which to accept" is what actually gets done. The boxed steps are the steps the researcher must perform. The null hypothesis appears automatically with the hypothesis, and the statistical software performs all the other steps.

The following guidelines are meant to help you choose a statistic, or understand why someone else chose it, but they contain only very commonly used and straightforward statistics. Many, many, many statistics, such as measures of "informative missingness," are appropriate for slightly different situations, most of which only arise if a sample size is small or sampling could not be completed as planned. So, make sure that you have a big sample size (rule of thumb $n \geq 30$ is a large sample size). You can scan the boldface examples to find data like yours will be, or you can read through the whole list to find the situation most like yours; it is a short list.

Descriptive Statistics

If your data are: numbers
Examples: **anything**
Statistics to choose from: FREQUENCY DISTRIBUTION, MEDIAN, IQR, OUTLIERS

If your data are: numbers that are normally distributed
Examples: characters that vary but are under selection pressure—**lengths, areas, volumes;** means of non-normal data—**mean age, mean height, mean fluorescence intensity**
Statistics to choose from: FREQUENCY DISTRIBUTION, MEAN, VARIANCE, STANDARD DEVIATION

Statistical Inference

If your data are: numbers that are normally distributed, and you know the standard deviation of the population

Examples: characters that vary but are under selection pressure—lengths, areas, volumes; means of non-normal data—mean FRET efficiency, mean fluorescence half-life

You can do the following: compare means, predict the population mean

Statistics to choose from: z TRANSFORMATION (COMPARE TO THE NORMAL CURVE)

If your data are: numbers that are normally distributed, but you do not know the standard deviation of the population

Examples: lengths, weights, volumes, means

You can do the following: compare means, compare the experimental to the control

Statistics to choose from: t-TEST (COMPARE TO t DISTRIBUTION)

If your data are: ratios of numbers, spreads of sample data

Examples: allele frequencies, variance, proportion of experimentals affected

You can do the following: compare observed to expected ratio, compare variances

Statistics to choose from: χ^2 TEST (COMPARE TO χ^2 DISTRIBUTION)

If your data are: numbers that are normally distributed and can be tabulated using only + and −, that is, success or failure, yes or no.

Examples: presence or absence of effect of mRNA injection, toxicity of treatment

You can do the following: compare the number of treated individuals showing the phenotype to the number of untreated individuals showing the phenotype.

Statistics to choose from: BINOMIAL COEFFICIENT (COMPARE TO BINOMIAL DISTRIBUTION)

If your data are: measurements that vary with changes in some second variable such as time

Examples: PCR product as a function of T_m, reaction rate, cell proliferation, dispersion of units, degree of effect as a function of dosage, cell area as a function of substrate area

You can do the following: determine the probability that the change in x *causes* the change in y

Statistics to choose from: r^2 (CORRELATION COEFFICIENT)

DESCRIBING DATA: DESCRIPTIVE STATISTICS

Frequency Distributions, Medians, and Quantiles

The first thing to do with your data is to make a picture. Start by having your statistics program create a frequency distribution, as shown below:

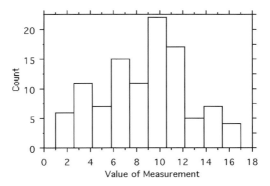

A frequency distribution is a graph with possible values of the measurement on the *x* axis, and the number of times that value occurred in the data set (i.e., the sample) on the *y* axis. The graph should be a histogram, with the area of the bar proportional to the value of the count, as above. (**WARNING:** Your program may offer other graphics. Do not use them because features such as three-dimensional columns can be visually distracting and even possibly deceptive.) The *x* axis is divided into intervals; the intervals are determined by taking the highest value and the lowest value in the dataset and then evenly dividing that range. Usually there are 10 intervals of equal width, but you can change that number. For example, you would want fewer intervals if any of the intervals were empty.

You can report these data by reproducing the picture, or you can sum up the information in the histogram with a few well-chosen numbers. Your program will find these numbers for you if you ask it to calculate Descriptive Statistics. The output will look something like this:

Count	**105**
Minimum	**1.0**
Maximum	**17**
Range	**16**
Median	**9.0**
IQR	**5.0**
Mean	8.6
Variance	14
Standard deviation	3.8

The program will probably give you more numbers than are shown here. The statistics that are not listed here are not discussed in this book. The statistics printed in boldface can be used to meaningfully describe any dataset; the others are applicable only to data that are normally distributed.

Talking about Descriptive Statistics

Count: The number of measurements, or the sample size.

Minimum and Maximum: The minimum is the lowest value measured; the maximum is the highest value measured.

Range: The range is the difference between the smallest and largest values, i.e., the maximum minus the minimum.

The Median: The median is the middle of the data in the sense that it divides the list, which has been sorted according to magnitude of the value, into two groups with equal numbers of points. The data in the larger half have values larger than the median; the data in the smaller half have values that are lower. The median may be all you want to know. If, for example, you are studying the toxicity of a drug, you might only need to know the dose that results in 50% of the subjects dying, a common measure called the LD_{50}. The LD_{50} is the median.

Interquartile Range (IQR): To give a little more information about the spread of the data around the median (which divides the data in two), the sorted data are divided into more categories, called quantiles, with even numbers of data points (as opposed to even intervals of values) in each quantile. If the data are divided into four quantiles, the quantiles are called quartiles. If the data are divided up into 100 quantiles, the quantiles are called percentiles.

The value of the $(1n/4)$th point, where n is the count (the sample size), is the first quartile, Q_1. The value of the $(2n/4)$th point is the second quartile, Q_2 (which is the median). The value of the $(3n/4)$th point is the third quartile, Q_3. The IQR is $Q_3 - Q_1$. The IQR can tell you if there are any outliers in your data.

Outlier: An outlier is a value that is so far away from the other values, it will have undue influence on certain descriptive statistics. If the value of a point is less than $Q_1 - 1.5 \times IQR$ or greater than $Q_3 + 1.5 \times IQR$, that point is an outlier.

Box Plots

The median, IQR, and outliers can be summed up in a picture, called a box plot, or a box and whiskers plot:

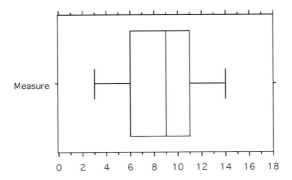

The box runs from Q_1 to Q_3; the line in the box is at the median; the "whiskers" extend to the 10th and 90th percentiles (sometimes the whiskers extend to the farthest points that are not outliers). If there were outliers in these data, they would be represented individually, as dots aligned with the whiskers.

The box plot may not seem to be a huge improvement over the frequency distribution picture. But, if you are trying to illustrate a number of frequency distributions, for example, if you want to make a quick decision about which protocol gives the most product or which protocol gives the most consistent amount of product, the statistics program can turn the data from each set of trials into a box plot, and put them all on the same axis. This plot gives lots of information, quick visual comparison, and only one figure.

Descriptive statistics can be used to describe the midpoint and spread of *any* dataset that can be represented by a frequency distribution. The other two descriptive statistics covered here are the mean and the standard deviation. Like the median and the IQR, these indicate where the middle of the data is and how the rest of the data are spread out around that midpoint. Mean and standard deviations are probably the most common descriptive statistics you'll see, and they are great ones; but they are not for every kind of data. Specifically, if the frequency distribution of the sample is not shaped essentially like a mound (high in the middle, low on either end, symmetrical), then the mean and the standard deviation are not as meaningful as they should be. For the mean and the standard deviation to live up to their potentials, the sample they are being used to describe has to be normally distributed.

Normal (Gaussian) Distribution; a Bell Curve

To say a sample is normally distributed is to say something about the shape of the frequency distribution. A normal distribution looks like a cross section through a pile of sand. A normal distribution is a frequency distribution that has its highest point more or less in the middle, is more or less symmetrical about the middle, and has no outliers. (The formal definition is a little more rigorous.)

Looking for Normality

Your statistics program probably offers a number of ways to judge whether the data are normal. One graphical way is shown here:

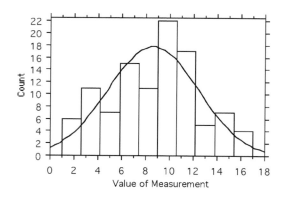

Superimposed over the frequency distribution is the outline of a normal version of the data; i.e., the curve is the outline of an imaginary normal distribution with the same mean and standard deviation as the sample. The software creates that ideal version of the sample by taking each value actually measured and transforming it to the value it would have been ("the ideal normal value") had the data actually been perfectly normal. In many cases, a glance at the comparison between the sample and the perfectly normal version of the sample is sufficient. If there is any doubt, however, the program can do a test, called a goodness-of-fit test, that will quantify whether the sample is "close enough" to normal.

Luckily, most measurements of things in the natural world are distributed normally. You can think about why in the following way: Most of the values of a character will be clustered around some "optimal" measure (the mean), due to selection. Because of genetic variation, environmental fluctuation, etc., there will always be a range (the spread); but the bigger the divergence from optimal, in either direction, the fewer examples there will be, because selection is removing them from the population. Thus, the frequency distribution of character values is symmetrical about the middle and has no outliers.

Testing for Normality

One test for normality is the Kolmogorov-Smirnov (sometimes abbreviated K-S) goodness-of-fit test for normality. This test indicates how well the data "fit" the ideal normal version of the data. If you use this test, the program will report some or all of the following results:

DF (Degrees of freedom)	2
Count, actual	105
Count, ideal normal	105
Maximum difference	0.067
Chi square (χ^2)	0.933
p-value	>0.999

The critical number in this table is the p-value. This p-value is the probability that you will be correct if you accept the null hypothesis that the distribution of the sample is the same as a normal distribution (i.e., the sampled population is normally distributed). In this case, it is very big (close to 1.000), so it is pretty safe to say that the sample is normally distributed. As long as $p \geq 0.95$, you are safe in saying the distribution is normal.

In the unlikely scenario that the data are not normally distributed, **STOP HERE**. You need what are called nonparametric, or distribution-free, statistics and they are not covered in this book. Most statistics texts have sections on nonparametric tests, such as the Mann-Whitney test, and there are texts devoted exclusively to nonparametric statistics. If the data are normally distributed, read on.

If the data are normally distributed, they are described using the mean and the standard deviation. The reason is that all normal distributions have the following characteristic: 95% of the values fall within 1.96 standard deviations of the mean; 99% of the values fall within 2.58 standard deviations of the mean. This turns out to be a profound piece of knowledge that will save you time and money. It also means that the data can be described with two numbers that convey a great deal of information.

INTERPRETING DATA: STATISTICAL INFERENCE

This section discusses some common questions:

- If these are my sample descriptive statistics, what can I say about the population, and how sure should I be?

- Is there a difference between the means of these two samples, i.e., how likely am I to be wrong if I claim that these samples are different?

- Is there a relationship between my variables? How much influence does one have on the other?

Inferring the Mean of the Population from the Description of the Sample

Statistical inference can be used to estimate values of characteristics of populations from a single sample of individuals taken from that population, and it can put a number on how confident you can be about that estimate. This is so because statistical inference is based on the laws of probability. Probability is not covered in this book, beyond the brief ideas presented in the introduction to this chapter, because this book assumes that you just want to know enough to be able to understand other people's statistics and, maybe, to calculate a few of your own. Textbooks and courses *do* cover probability; so if you are interested, the explanation is available. For now be assured, probability does explain inference.

To understand the following example, think about it from an omniscient perspective. Suppose you know the mean and standard deviation of the weight of a normally distributed population of the 10,362 tadpoles in a pond. Imagine that someone comes along and takes a sample: A scientist randomly and without bias grabs 250 tadpoles, measures their weight, and calculates a mean and standard deviation. Now imagine that another scientist grabs 250 tadpoles and calculates their mean and standard deviation, and someone else does it, etc., etc., etc. By chance, each of the scientists is more likely to pick individuals whose weight is near the mean, simply because there *are* more individuals whose weight is close to the mean. To understand why, think about the shape of a normal distribution, i.e., the actual distribution of weights in the pond. Because the actual weights are normally distributed, there are more individuals with weights that are close to the mean weight, and fewer that are very different from the mean. Nevertheless, after they have finished sampling, all of the scientists will calculate different means and standard deviations.

Most of those means will be closer to, but some will be farther from, the actual mean; in other words, there will be a distribution of sample means. Wonderfully, this distribution of the sample means will definitely be normal (this fantastic result is known as the Central Limit Theorem), and the mean of the sample means is a very good estimate of the mean of the population. If you do a lot of samples and then calculate the mean of the sample means, you will have a good estimate of the population mean. But, it turns out you don't have to do that. You can use the knowledge gained from the omniscient perspective to determine how good an estimate you can get from a single sample. And, wonderfully, how good the estimate is does not depend on the population size, only on the sample size.

Estimating the Population Mean
if the Sample Size Is Large

As a rule of thumb, $n \geq 30$ is a large sample size. Because the distribution of the sample means *would be normal* if you had them all, you already know something about the individual points within that distribution. That is, 95% of them lie within 1.96 population standard deviations of the population mean, and 99% of them lie within 2.58 population standard deviations of the population mean. So, even if you only have one sample mean and its standard deviation, denoted $\bar{x}$ and s, there is a 95% probability that $\bar{x}$ is within 1.96 population standard deviations of the population mean, and a 99% probability that it is within 2.58 population standard deviations of the population mean. Because you don't have the population standard deviation, you use the sample standard deviation, s, as an estimate. It's not perfect, but it is often the best there is. Because you only have a "best there is," and not an optimal value, you compare your data to a t-distribution (instead of a normal distribution) and you've still got your answer.

So, to estimate the population mean, you calculate (i.e., the computer calculates) the sample mean and standard deviation; then it calculates (by comparing those numbers to the t-distribution) the range of values that could be the actual mean, based on the fact that the sample mean has a 95% probability of lying within 1.96 standard deviations of the actual mean. The computer can calculate the range of values—the smallest and largest values the population mean might be; these are the 95% confidence limits. Or, it can calculate the probability that you would be wrong if you claimed your sample mean represents the population mean (the p-value).

Confidence limits

Confidence limits are the best way to express the value you think is the population mean—you indicate the value of the sample mean and then you indicate the probability that the true mean is ± the confidence limit; this is the confidence interval. You can also have the computer give you different confidence limits, based on whether you want to be 95% confident or 99% confident or whatever confidence level you choose. To tell the computer what confidence limits you want, you must enter an α value.

α Level

The α level is, basically, the probability of being wrong. If $\alpha = 0.05$, that means there is a 5% chance of being wrong, which is a 95% chance of being right, which is a 95% confidence limit. To convert from confidence limit to α, use: $\alpha = 1 - (0.01 \times \%)$; therefore, for 95%, $\alpha = 1 - 0.95 = 0.05$.

For the following example, I input a column of 200 numbers between 17 and 37, then the statistics program was asked for the mean, the standard deviation, and the 95% and 99% confidence limits. For the latter two numbers, the actual input was $\alpha = 0.05$ and then $\alpha = 0.01$. This was the output:

Mean = 24.73
Standard deviation of the mean = 4.19
Confidence ($\alpha = 0.05$) = 0.58
Confidence ($\alpha = 0.01$) = 0.76

What it means:

Sample mean ($\bar{x}$) = 24.73
Sample standard deviation (s) = 4.19
95% Confidence interval for the mean of the population = 24.73 ± 0.58
99% Confidence interval for the mean of the population = 24.73 ± 0.76

It also says that there is a 5% chance of being wrong if I claim that the true mean is between 24.15 and 25.31, and a 1% chance of being wrong if I claim that the true mean is between 23.97 and 25.49.

The smaller the confidence limits, the better you believe your estimate to be. To make them smaller, you increase the sample size. To get a sense of why, think back to the pond full of tadpoles for a moment. What would happen if one scientist measured only two tadpoles, while another scientist measured 200. Clearly, the mean of the 200 measurements is much more likely to be closer to the mean of the population. So, if every sample is large, the estimate of the mean of the population will be better. This is the same thing as saying the larger the sample size, the smaller the sampling error. It turns out that the confidence intervals depend on the square root of the sample size. Here are the formulas for confidence limits:

$$1.96 \times \frac{s}{\sqrt{n}}$$ for 95% confidence limits for the mean if n is large

$$2.58 \times \frac{s}{\sqrt{n}}$$ for 99% confidence limits for the mean if n is large

$\quad s$ = Standard deviation of the sample
$\quad n$ = (Large) sample size

The numbers 1.96 and 2.58 are called the z scores, or normal deviates. They are chosen to be appropriate to the confidence limit desired.

If your sample size is small (rule of thumb: $n \leq 30$ is small), you cannot assume that the distribution of the sample means will be normal. You can, however, assume something else about the distribution: It will be what is called a t-distribution. The difference is that this distribution is more spread out; so a 95% confidence interval will be bigger than it would be if you had a large sample size. For a small sample size, you need to use the "critical value of t" in place of the z score. Unlike the z score, which is always the same, critical value of t depends on the sample size.

For the following example, one column of 25 numbers between 17 and 37 was used as the sample to calculate the population mean from a small sample. Here are the statistics:

Sample mean = 24.79
Sample standard deviation = 4.28
95% Confidence limit = 1.77
99% Confidence limit = 2.40

The confidence limits come from the following calculation:

$$2.06 \times \frac{s}{\sqrt{n}}$$ for 95% confidence limits for the mean if n = 25

$$2.80 \times \frac{s}{\sqrt{n}}$$ for 99% confidence limits for the mean if n = 25

The 2.06 and 2.80 come from a table of critical values of the t distribution. For $\alpha = 0.05$ and 24 degrees of freedom, the computer returned a critical value of 2.06; for $\alpha = 0.01$ and 24 degrees of freedom, it returned 2.80. I used 24 degrees of freedom because I was calculating confidence limits for a mean, and for a mean, the degrees of freedom equals the sample size minus one, i.e., $n - 1$. So, if the sample size were different, those two numbers, the critical values of t, would be different.

A Note on Sample Size

The large sample size indicated that there is a 95% chance that the population mean is between 24.12 and 25.28. The small sample size indicated that there is a 95% chance that the population mean is between 23.02 and 26.56. If the latter is narrow enough, 175 measurements can be skipped. When in doubt, or if you wish to choose one way to always do it, use the *t*-distribution; the limits will be larger, but you can be sure you are reporting an accurate confidence interval.

A Note on Built-in Functions in Statistics Programs

Be careful—In the program I used for the examples above, the function that automatically calculates confidence intervals assumes a large sample size, and so it uses the normal deviates. Therefore, I could not use it when looking at the small sample size. That was not obvious from the command; I had to check to find out. This may happen to you. Again, when in doubt, use your statistics package to tell you the mean, the standard deviation, and the critical value of the *t*-distribution, and then do the calculation on a calculator.

Making Comparisons

Often you want to compare two data sets with each other to decide if they are different. You cannot, however, just look at the two sets of descriptive statistics and claim, for example, that the mean of sample A is different or not different from the mean of sample B. You have to report the probability that you are wrong (and thus the probability that you are right). That probability is summed up in the *p*-value. If the probability that you are wrong about your conclusion is less than a particular value, you can claim that you are certain about your conclusion. That raises the question "how certain is certain enough to claim that you have uncovered a truth?" The level of certainty required is determined by tradition and, by tradition, 95% certain is usually considered certain enough. If you are 95% certain that you are correct, that means there is a 5% probability that you are incorrect. If the probability that you are incorrect is 5% or smaller, then your conclusion is said to be "statistically significant."

The 5% probability that you are wrong is what you report, and you report it by reporting the *p*-value. The *p*-value is the number that says how likely it is that you have drawn an incorrect conclusion; that is why the smaller the *p*-value, the better.

Practicalities

There is a basic pattern to the procedures for using statistics to evaluate claims about comparisons. Here is generally what to do.

METHOD

1. Make sure the data satisfy the assumptions of the procedure. (Put another way: know what procedure you will be using before you collect the data and then make sure you collect the data correctly.)

2. Decide what level of probability you will consider significant (or look it up in the journal to which you want to submit the work; usually it is $p \leq 0.05$ or $p \leq 0.01$).

3. Have the computer calculate the test statistic, and compare the statistic to the distribution to get the p-value. If the p-value is less than or equal to the cutoff, reject the null hypothesis that there is no difference and claim that your two samples are different.

4. Report *both* the value of the statistic and the p-value.

This was summarized in a figure earlier; for convenience, it is reproduced here:

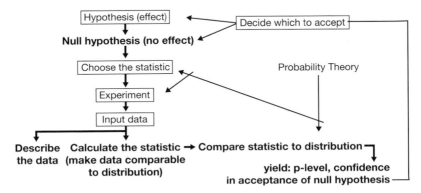

An example of how this scenario works follows.

☑ **Example: Comparing Two Means: The t-test, aka Student's t**

You want to know if the length of your mutant yeast cells is different from the wild-type cell length. Around the loop step-by-step:

1. **Hypothesis.** You believe your mutants are shorter than wild type; therefore, the null hypothesis is that there is no difference in length. You will be comparing two numbers.

2. **Choose the Statistic.** You will not know the actual spread of the measurements in the population, because no one has ever measured this mutant before. You will need to compare your data to a *t*-distribution; so plan to do a *t*-test. To do a *t*-test, you need normal data. The distribution of lengths is probably normal, but you suddenly remember that *means* are *always* normal (the Central Limit Theorem); so you decide to compare the *mean* length of a sample from the mutants to a *mean* length of a sample from the wild types.

3. **Experiment.** You randomly and independently select 60 cells to measure from each of the populations (mutants and wild types), and you measure them.

4. **Input Data.** You put the 60 values from the wild-type cells into the first column of the spreadsheet; you put the 60 values from the mutant into the second column of the spreadsheet.

5. You use the computer to calculate the mean and standard deviation of each column, so that you can describe the data. It tells you that, for the wild type, length is 22.2 ± 1.4 µm and for the mutant, length is 17.6 ± 1.6 µm. Just for laughs, you have the computer create a frequency distribution of each sample; it looks like the mutants are shorter. You press on.

6. You try to have the computer calculate the value of *t* for the sample and then compare that value to the *t*-distribution, but you discover that you need to input the degrees of freedom and whether the distribution should be one- or two-tailed. Sample size is 60, degrees of freedom for the mean are $n - 1$, and so DF = 59. You believe the mutants are shorter (not just different); so one tail. The computer indicates that $t = 2.89$ and $p = 0.003$. You have a 0.3% probability of being wrong if you reject the null hypothesis, and so you can have >99% confidence that the two means are different.

7. **Decide Which to Accept.** Accept your hypothesis. Claim confidently that the mutants are shorter. When you publish, be sure to include both $t = 2.89$ and $p = 0.003$; just reporting the p level isn't good enough.

STATISTICAL INFERENCE: RELATIONSHIPS

Correlation

...is NOT causation. Yes, it is hard to believe, but contrary to the implications of much political rhetoric and most advertisements, observing a correlation is not the same as proving causation. There are many great illustrations of this; one of the best is Mark Clifton's "The Dread Tomato Addiction," which correctly points out, as evidence that tomatoes are evil, that there is a *perfect* correlation between being dead and having eaten tomatoes during the first half of the 19th century. In fact, to show causation, you first must establish correlation, but then you must prove necessity and sufficiency, which is much harder. What follows deals with the simple task of quantifying your confidence in the ability of a line to describe the correlation between variables.

The most common measure of correlation is the coefficient of correlation. The correlation coefficient, r, is related to the idea of covariance. Covariance, as the name suggests, is the idea that it may mean something if an increase in the variable studied always happens when there is an increase in the independent variable, or if an increase in one is always accompanied by a decrease in the other—the values don't just vary, they co-vary. A statistic called the covariance addresses this. The problem is that the covariance depends on the scale of the measurement used. This is potentially a problem: If you are measuring growth (height change over time) in centimeters, and someone else in the lab is measuring growth in millimeters, your covariances will differ by a factor of ten, even if the relationships you measured were actually the same. So, a unit-independent measure of covariance was devised: the correlation coefficient. Mathematically, the correlation coefficient is just the covariance divided by the standard deviations of the two variables; this eliminates the scale problem and gives you the correlation coefficient, which is always between 0 and 1.

The circumstances under which you would want to calculate a correlation coefficient are as follows. (This will be discussed as if you are done with the experiment. Do not forget that in reality you will have decided before you started that you were going to analyze the data this way; thus, you were able to avoid mistakes that would invalidate the analysis.) You have measured the concentration of a chemical at different time points, or at different temperatures, or as you vary some other quantifiable variable. You have made these measurements randomly, i.e., in no particular order. For example, you did *not* take a reading at the lowest temperature, then the next highest, then the next, then the next, etc.; you know enough about sampling to have varied the temperature according to some random sequence. (In this particular case, it is so that you will know it is temperature and not time with which your data are correlated.) You then graphed concentration of chemical (*y* axis) as a function of temperature (or time, or whatever; *x* axis). There is obviously a linear increase in concentration with increasing temperature, but you need a number to report. So, you have the computer find the linear regression line; you may also have to indicate whether (0,0) is a datum. What the computer does in response is to find a line that minimizes the distance between the line itself and the actual data: It finds the line with the "best fit" to the actual data; any other line would be less faithful. The next step is to find a number that summarizes if that best-fit line is actually any good. The better it is, the smaller the distance between it and the data. So, you calculate a number, the correlation coefficient, that tells you if that distance is small or large. Unlike *p*-values, which you usually want to be small, you want *r* to be as large as possible, i.e., as close to 1.000 as possible.

Although it is not covered here, it is possible to use nonlinear regression if the relationship between the variables is nonlinear. Check with the nearest statistician.

REPORTING NUMERICAL DATA

There are different ways to report quantitative data. Which you choose depends on which will best communicate the information. That depends on your audience and on the format of the communication. What is best in a published paper may not be what is best for a talk. The following includes descriptions of common ways to report numerical data and advice on when to use what.

Scale Bars

Scale bars are important pieces of numerical data that you may have to work with even if you studiously avoid all other numbers. A scale bar is critical because you must indicate the size of the objects in a figure.

Stating the magnification of a figure is meaningless now that most copy machines can shrink and enlarge, and now that figures can be altered electronically. Scale bars, however, will grow and shrink right along with the rest of the figure; so they will remain meaningful, no matter what happens. A scale bar should be a useful size. For example, a bar longer than 25% of the picture is too big; a bar that is smaller than anything of interest in the micrograph is too small.

Many types of digital imagers will automatically put a scale bar on an image. If you create an image using something else, you will have to add the scale bar by hand. To create a scale bar by hand, you have to make an image of a ruler or a stage micrometer that is magnified the same amount as the image. If the image is electronic, graphics software, such as Photoshop can be used to add a line electronically; but, you will still need a picture of a ruler. If you are using graphics software, you must be careful that the resolution of the image of the ruler is the same as the resolution of the image you are labeling. If it isn't, cutting and pasting may alter the relative lengths and the scale bar will be wrong. The following is a method for using rub-on lines to create scale bars on photographs.

METHOD   **Scale Bar**

1. Determine about the right length for the scale bar.

2. On the micrograph of the micrometer (or photo of the ruler), choose two lines that are about the right distance apart.

3. Note the distance marked off by the two lines (see Chapter 3). Use those two lines to mark the beginning and end of the rub-on line. To mark the line, either make an inconspicuous mark on the micrograph or use a razor to gently trim the rub-on line to the proper length directly on the dry transfer sheet (if you use the cutting technique, be careful not to cut the plastic backing).

4. To apply the line, position the plastic sheet over the figure (tacky side against the figure, shiny side facing you) so that a line lines up with your mark, or so that the already trimmed line is near an edge (usually the lower right) and does not obscure anything of interest.

5. With a dull pencil, rub the length of line you wish to transfer. It will stick to the micrograph.

6. Gently lift the plastic away and check the result. If you make a mistake, dry transfers can be removed using Scotch Tape. Pat the sticky

side of the tape to the line you want to remove; it should come off with two or three pats.

7. After you have put down the line, you must indicate what length it represents. Go back to where you noted the distance between the two lines (step 3). That is the distance represented by the line.

8. Using dry transfer letters, write the length, either just above or just below the bar.

Graphs

A graph is one way to represent numerical data. A table is another. Tables are preferable if you have a few data points; a graph is preferable if you have many. The point of graphing data is to see whether there is a relationship between the variables and whether that relationship can be described by an equation. If there is a mathematical relationship between the variables, you can use an equation to describe the data, you can use the equation to predict the results of experiments that you have not done, and you can, in the future, measure any one variable and immediately know the value of the other. Describing a relationship using an equation can save you a lot of work.

Determining what the equation is requires a priori knowledge and/or a curve-fitting algorithm. Determining how well an equation describes the data requires statistics (see the discussion of correlation coefficient, above).

Conventions of graphing

- Most graphs have two axes, x and y. If there is a third, it is designated z.

- The origin is at (0,0); magnitudes increase with distance from the origin.

- The variable on the x axis is the independent variable; i.e., the variable that the researcher controls. Time also goes on the x axis.

- The variable on the y axis is the dependent variable, i.e., the variable the researcher is measuring and which varies as a result of a variation in x. The value of y "depends" on the value of x.

Lines

- The general equation for any line is $y = mx + b$, where m is the slope and b is the y intercept. The x intercept is $-b/m$.

- The x axis is the graph of the equation $y = 0$.
- The y axis is the graph of the equation $x = 0$.
- If the slope of the line is positive, the dependent variable is said to vary positively, or directly, with the independent variable. If the slope of the line is negative, the dependent variable is said to vary negatively, or inversely, with the independent variable.

Transforming data

Graphing raw data does not always make relationships evident. Data are frequently manipulated mathematically so that relationships will become obvious. For example, because lines are much easier to interpret than curves are, data that are curvy can be mathematically tweaked so that a line will be a good descriptor. To transform data means to perform an operation on each of the points. Common transformations include:

- Taking the log of the data. This makes data that are very spread out at high values and closer together at lower values space out more evenly over their entire range.

- Taking the reciprocal of the data. This turns a certain kind of curve into a line. (For an example, see Lineweaver-Burke plots, Chapter 6.)

  Data are also transformed to increase clarity. This usually means tweaking so that the range of values of the dependent variable becomes 0–1 (0–100%), or so that the dependent variable becomes dimensionless (this makes it simpler to compare the data with results from other experiments). Dividing the value of each point by a particular constant is called normalizing. The constant is frequently the value of the control, the value of the mean of the data, or the highest value measured. Either or both variables can be transformed. If the variable is transformed, its units must be transformed as well.

WORDS, TABLES, OR GRAPHS?

Deciding on a Format When Writing a Research Paper

When deciding whether to publish numerical results using words, tables, or graphs, the important things to keep in mind are ink and clarity. Never waste ink, never sacrifice clarity. Use words if you are reporting a few numbers that are not easily confused. If you are reporting the number of specimens you measured (the sample size), you do not need to draw a picture; the datum is a single number, whose meaning is perfectly clear from the phrase "$n = 426$." Don't waste ink producing a graph or making an icon of your specimen and showing 426 of them. Likewise, if you are reporting the mean values of measurements on two different populations, say "sample A had a mean depth of 30 meters, whereas sample B had a mean depth of 25 meters." Don't devote space and ink to a table with this information in it; it is obvious that A is deeper than B. In other words, if you are reporting a few numbers, and your reader is not likely to get confused about what they mean, use words.

Use a table if (1) you are reporting a list of numbers that is too long for the average reader to keep it all straight, (2) there is a simple pattern in the numbers you wish to emphasize, or (3) only one of the variables is numeric. For example, the average number of blossoms on plants fed six different fertilizers is best reported in a table, in either ascending or descending order of blossom number. A graph would insinuate a numeric pattern in the fertilizers (an axis connotes differences in magnitude, not just differences in identity), and it would be unlikely to reveal more information than a table. Words would not reveal the trend very well and would probably not be memorable. A table is not too much, and not too little. A table would be just right.

Use a graph when you wish to illustrate the relationship between two numeric variables. If the list of numbers is short, indicating that you should use a table, but the trend in values is complicated or subtle, you should probably use a graph. For example, if you are measuring an increase in chemical concentration at six time points, and the increase is slow at first, then fast, then slow again, a graph will be a much better illustration than a table; your reader will be able to see immediately that there is an S shape to the curve, and this gives the reader important information about the process.

Deciding on a Format When Preparing a Presentation

When deciding whether to make a slide of numerical results using words, tables, or a graph, the important things to keep in mind are visual impact and clarity. Never distract and never sacrifice clarity; feel free to waste ink.

Use words when illustrating numerical results *only* as a last resort. Pictures are always more compelling (and memorable) than words. If you would use words in a manuscript, chances are you don't need a visual at all. A slide saying "$n = 426$" is a waste of time. If it is important that your audience know the number, add it to the slide showing the data or the results, or just say it out loud (however, see discussion on using pictures below).

Use words instead of (or at least in addition to) symbols if you are showing an equation. Word equations usually help to make the meaning of the equation clear. Also, the audience will probably not remember what each symbol stands for, but they will remember how to read. Another practical reason for using words instead of symbols is that, unless you are speaking to an audience composed entirely of math-comfortable and math-interested listeners, an equation slide is likely to cause much of the audience to start drifting away. Word equations are a nice compromise if you need to explain an equation. As with any slide, however, do not overload. If the equation is long and complicated, break it up into intuitively explainable pieces, devote a slide to each piece, go over those individually, then reassemble the entire word equation in a summary slide.

Use tables when you have six or fewer numbers and there is no particular pattern you wish to illustrate. Those six numbers can be one column with six rows, two columns with three rows, etc. Obviously, the number six is not an absolute cutoff; but, given the amount of time an audience has to absorb the information on a slide, six different numbers, with their attendant comparisons, is sufficiently close to the limit of comprehensibility.

Use graphs liberally when giving a talk, and write the conclusion in words somewhere on the slide. The axes should be labeled clearly with both numbers and units. More slides, each of which has a single graph that illustrates a single facet of the data, are much better than fewer slides, each of which is dense with information. Don't forget that the audience is performing the variety of complicated mental tasks involved in learning something based on a single, rapid, audiovisual presentation. Don't tax them further by making them work to understand each slide. You are more likely to get your points across if you present those points with the audience's task in mind. Finally, no matter how clearly labeled they are, *always* begin your description of a graph by explaining what the axes represent.

Use pictures of numerical data liberally when giving a talk. Graphics you would never waste ink on in a manuscript can be very useful in a talk. For example, if it is *important* that the sample size was 426, showing a slide with 426 icons can nicely reinforce the magnitude and importance of the number. Another good example is color-coding protein sequences that are being compared, using contrasting colors to illustrate the positions of changes. A table listing the numerical positions of changed amino acids, and what those changes are, would contain exactly the same information, and it would save ink, but the audience would almost certainly fall asleep during your explanation.

RESOURCES

Note on Statistics References

If you search the Web for help with statistics, try the keywords "statistical infer-ence" or "statistical test" or the name of the particular test; just "statistics" will lead you to pages and pages of statistics, the numbers, not Statistics the process.

Choosing a Statistical Test

http://www.graphpad.com/www/book/Choose.htm

http://bmj.com/collections/statsbk/13.shtml

http://www.ats.ucla.edu/stat/mult_pkg/whatstat/default.htm

Reporting Numbers

Tufte E.R. 2001. *The visual display of quantitative information*, 2nd edition. Graphics Press, Cheshire, Connecticut.

Statistics Texts

Box G.E.P., Hunter W.G., and Hunter J.S. 1978. *Statistics for experimenters: An intro-duction to design, data analysis, and model building*. John Wiley & Sons, New York.

Zar J.H. 1984. *Biostatistical analysis*, 2nd edition. Prentice-Hall, Englewood Cliffs, New Jersey

Text

Graunt J. 1665. Letter (1662) to John Lord Roberts in *Natural and political obser-vations mentioned in a following index, and made upon the bills of mortality*. J. Martyn and J. Allestry, London. (Quoted in Weaver J.H. [2001] *Conquering statistics: Numbers without the crunch*. Perseus Publishing, Cambridge, Massachusetts.)

Tomato Addiction

http://members.tripod.com/~mmonroe/tomato/addiction.html

Understanding Statistics

Gonick L. and Smith W. 1993. *The cartoon guide to statistics*. HarperPerennial, New York.

Huff D. 1954. *How to lie with statistics*. W.W. Norton, New York.

Weaver J.H. 2001. *Conquering statistics: Numbers without the crunch*. Perseus Publishing, Cambridge, Massachusetts.

CHAPTER 8

Reference Tables and Equations

GREEK SYMBOLS

Greek		Computer Key	(United States Pronunciation) and Some Standard Uses of UPPERCASE and lowercase Symbols
A α	alpha	A a	
B β	beta	B b	
Γ γ	gamma	G g	micrograms
Δ δ	delta	D d	CHANGE IN; change in
E ε	epsilon	E e	
Z ζ	zeta	Z z	(zate-ah)
H η	eta	H h	(ate-ah) symbol for viscosity
Θ θ	theta	Q q	(thate-ah)
I ι	iota	I i	
K κ	kappa	K k	
Λ λ	lambda	L l	microliters; symbol for wavelength
M μ	mu	M m	('myoo') micro-; symbol for population mean
N ν	nu	N n	symbol for frequency
Ξ ξ	xi	X x	(zi or k-sigh)
O o	omicron	O o	
Π π	pi	P p	3.1416; the circumference of a circle divided by its diameter
P ρ	rho	R r	symbol for density
Σ σ	sigma	S s	SUM; symbol for population standard deviation
T τ	tau	T t	
Y υ	upsilon	U u	
Φ φ	phi	F f	(fee)
X χ	chi	C c	(ki)
Ψ ψ	psi	Y y	(sigh)
Ω ω	omega	W w	SYMBOL FOR OHMS; symbol for angular frequency

PREFIXES FOR UNITS

Prefix	Abbreviation	Multiplier
yotta-	Y	10^{24}
zetta-	Z	10^{21}
exa-	E	10^{18}
peta-	P	10^{15}
tera-	T	10^{12}
giga-	G	10^{9}
mega-	M	10^{6}
kilo-	k	10^{3}
hecto-	h	10^{2}
deca-	da	10^{1}
deci-	d	10^{-1}
centi-	c	10^{-2}
milli-	m	10^{-3}
micro-	μ	10^{-6}
nano-	n	10^{-9}
pico-	p	10^{-12}
femto-	f	10^{-15}
atto-	a	10^{-18}
zepto-	z	10^{-21}
yocto-	y	10^{-24}

PREFIXES FOR NOMENCLATURE

Prefix	Multiplier
hemi-	1/2
mono-	1
di- bi- bis-	2
tri- tris-	3
tetra-	4
penta-	5
hexa-	6
hepta-	7
octa-	8
nona-	9
deca-	10

SYSTÈME INTERNATIONALE UNITS: SI UNITS

Property	Symbol for Property	Dimensions	SI Unit	Symbol for Unit	Symbol in Base Units
Length	l, d	**L**	**meter**	**m**	**m**
Mass	m	**M**	**kilogram**	**kg**	**kg**
Time	t, τ	**T**	**second**	**s**	**s**
Electric current	I	**A**	**ampere**	**A**	**A**
Temperature	T	Θ	**kelvin**	**K**	**K**
Amount of substance	n	**N**	**mole**	**mol**	**mol**
Luminous intensity	I_v	**J**	**candela**	**cd**	**cd**
Plane angle	α	–	radian	rad	$m\,m^{-1}$
Solid angle	Ω	–	steradian	sr	$m^2\,m^{-2}$
Area	A	L^2	meters squared	m^2	m^2
Volume	V	L^3	meters cubed	m^3	m^3
Volume		L^3	liter	l or L	$10^{-3}\,m^3$
Frequency	f	T^{-1}	hertz	Hz	s^{-1}
Radioactivity		T^{-1}	becquerel	Bq	s^{-1}
Rate, speed, velocity	U, v	$L\,T^{-1}$	meters/second	$m\,s^{-1}$	$m\,s^{-1}$
Angular velocity	ω	T^{-1}	radians/second	$rad\,s^{-1}$	$m\,m^{-1}\,s^{-1}$
Acceleration	a	$L\,T^{-2}$	meters/second squared	$m\,s^{-2}$	$m\,s^{-2}$
Molarity	M	$N\,L^{-3}$	mol/liter	M	$mol\,10^3\,m^{-3}$
Density	ρ	$M\,L^{-3}$	kilograms/ meter cubed	$kg\,m^{-3}$	$kg\,m^{-3}$
Concentration	c	$M\,L^{-3}$	kilograms/ liter	$kg\,L^{-1}$	$kg\,10^3\,m^{-3}$
Force, weight	F, w	$M\,L\,T^{-2}$	newton	N	$kg\,m\,s^{-2}$
Pressure	p	$M\,L^{-1}\,T^{-2}$	pascal	Pa; $N\,m^{-2}$	$kg\,m^{-1}\,s^{-2}$
Energy, work	E, W	$M\,L^2\,T^{-2}$	joule	J; N m	$kg\,m^2\,s^{-2}$
Power	P	$M\,L^2\,T^{-3}$	watt	W; $J\,s^{-1}$	$kg\,m^2\,s^{-3}$
Electrical charge	Q	$T\,A$	coulomb	C	$s\,A$
Electrical potential	V	$M\,L^2\,T^{-3}\,A^{-1}$	volt	V; $W\,A^{-1}$	$kg\,m^2\,s^{-3}\,A^{-1}$
Electrical resistance	R, Ω	$M\,L^2\,T^{-3}\,A^{-2}$	ohm	Ω; $V\,A^{-1}$	$kg\,m^2\,s^{-3}\,A^{-2}$
Electrical capacitance	C	$M^{-1}L^{-2}T^4A^2$	farad	F; $C\,V^{-1}$	$kg^{-1}\,m^{-2}\,s^4\,A^2$
Electrical field strength		$M\,L\,T^{-3}\,A^{-1}$	volts/meter	$V\,m^{-1}$	$kg\,m\,s^{-3}\,A^{-1}$
Inductance		$ML^2T^{-2}A^{-2}$	henry	H; $Wb\,A^{-1}$	$kg\,m^2\,s^{-2}\,A^{-2}$
Conductance		$M^{-1}L^{-2}T^3A^2$	siemen	S; $A\,V^{-1}$	$kg^{-1}\,m^{-2}\,s^3\,A^2$
Flux density		$MT^{-2}A^{-1}$	tesla	T; $Wb\,m^{-2}$	$kg\,s^{-2}\,A^{-1}$
Magnetic flux		$ML^2T^{-2}A^{-1}$	weber	Wb, V s	$kg\,s^{-2}\,A^{-1}\,m^2$
Luminous flux		J	lumen	lm; cd sr	cd
Illuminance		$L^{-2}\,J$	lux	lx; $lm\,m^{-2}$	m^{-2} cd
Luminance		$L^{-2}\,J$	candelas/meter squared	$cd\,m^{-2}$	m^{-2} cd
Heat capacity/entropy	S	$M\,L^2\,T^{-2}\,\Theta^{-1}$	joules/kelvin	$J\,K^{-1}$	$kg\,m^2\,s^{-2}\,K^{-1}$
Specific entropy		$L^2\,T^{-2}\,\Theta^{-1}$	joules/kilo- gram kelvin	$J\,kg^{-1}\,K^{-1}$	$m^2\,s^{-2}\,K^{-1}$
Thermal conductivity		$M\,L\,T^{-3}\,\Theta^{-1}$	watts/meter kelvin	$W\,kg^{-1}\,K^{-1}$	$kg\,m\,s^{-3}\,K^{-1}$

The seven fundamental units (SI base units) are assumed to be mutually independent; that is, none of the base units can be constructed by arranging any of the others. The fundamental units are shown in boldface letters.

OTHER UNITS

Property	Symbol for Property	Dimensions	Other Unit	Symbol for Unit	To Convert to SI
Plane angle	θ, α		degree	°	rad = degrees ÷ 57.3
Mass	m	M	dalton	D	kg = D ÷ (6.022142 × 10²⁶)
Temperature	T	Θ	° Celsius	°C	K = °C + 273.15
Temperature	T	Θ	° Fahrenheit	°F	K = $\frac{5}{9}$ °F + 255.37
Area	A	L^2	hectare	ha	m^2 = 0.0001 ha
Volume	V	L^3	cubic cm	cc	mL = cc
Force	F	$M\,L\,T^{-2}$	dyne	dyn	N = 10^5 dyne
Pressure	p	$M\,L^{-1}\,T^{-2}$	atmosphere	atm	Pa = 9.86926 × 10⁻⁶ atm
Energy, work	E, W	$M\,L^2\,T^{-2}$	calorie	cal	J = 0.239 cal
Energy, work	E, W	$M\,L^2\,T^{-2}$	erg	erg	J = 10^7 erg

"SQUARE ROOTS PI, NEGATIVE NUMBERS...
FRANKLY I HAVE A LOT OF DIFFICULTY
RELATING TO ALL THAT SYMBOLISM."

CONSTANTS

Constant	Symbol	Value
Angstrom	$\mathring{A}$	10^{-10} m
Atomic mass unit (amu)	u	$1.66053873(13) \times 10^{-27}$ kg
Avogadro's constant	N_A, L	$6.02214199(47) \times 10^{23}$ mol^{-1}
Avogadro's number	A	$6.02214199(47) \times 10^{23}$
Base of natural log (ln)	e	2.718 281828459
Boltzmann's constant	k	$1.3806503(24) \times 10^{-23}$ J K^{-1}
Complex numbers	i	$i = \sqrt{-1}$; $i^2 = -1$
Electron volt	eV	$1.602176462(63) \times 10^{-19}$ J
Elementary charge	e	$1.602176462(63) \times 10^{-19}$ C
Faraday's constant	F	$9.64853415(39) \times 10^4$ C mol^{-1}
Gravitational acceleration (standard)	g_n	9.80665 m s^{-2}
Gravitational constant	G	$6.673(10) \times 10^{-11}$ N m^2 kg^{-2}
Mass of a neutron	m_n	$1.67492716(13) \times 10^{-27}$ kg
Mass of a proton	m_p	$1.67262158(13) \times 10^{-27}$ kg
Mass of an electron	m_e	$9.10938188(72) \times 10^{-31}$ kg
Molar volume of ideal gas at STP	V_m	22.413996×10^{-3} m^3 mol^{-1}
Mole	mol	$6.02214199(47) \times 10^{23}$
Permeability of a vacuum (magnetic constant)	μ_o	$4\pi \times 10^{-7}$ N A^{-2}
Permitivity constant (electric constant)	ε_o	$8.854187817 \times 10^{-12}$ F m^{-1}
Pi	π	3.14159265358979323846264338327950
Planck's constant	h	$6.62606876(52) \times 10^{-34}$ J s
Planck's constant over 2π (h bar)	$\hbar$	$1.054571596(82) \times 10^{-34}$ J s
Rydberg's constant	R	10973731.568549(83) m^{-1}
Speed of light in a vacuum	c	2.99792458×10^8 m s^{-1}
Speed of sound (dry air, 0°K) (water, 20°K)	C	331.45 m s^{-1} 1470 m s^{-1}
Standard temperature and pressure	STP	273.15°K and 101.325 kPa
Stefan-Boltzmann constant	σ	$5.670400(40) \times 10^{-8}$ W m^{-2} K^{-4}
Universal or molar gas constant	R	$8.314472(15)$ J mol^{-1} K^{-1}

See http://physics.nist.gov/cuu/Constants/index.html.

The number in parentheses after the mantissa is the uncertainty in the last digits of the mantissa. For example, Avogadro's constant is written $6.02214199(47) \times 10^{23}$ mol^{-1}, which means $6.02214199 \times 10^{23} \pm 0.00000047 \times 10^{23}$ mol^{-1}. If no uncertainty is indicated, the number is exact.

USEFUL EQUATIONS FROM GEOMETRY, ALGEBRA, AND TRIGONOMETRY

Algebra

To find the distance d between two points (x_1, y_1), and (x_2, y_2), use the **Pythagorean theorem:**

$$d = \sqrt{(y_2 - y_1)^2 + (x_2 - x_1)^2}$$

To find the roots of a polynomial of the form $y = ax^2 + bx + c$, use the **Quadratic equation** $y = 0$, where:

$$x = \frac{-b \pm \sqrt{b^2 - 4ac}}{2a} \quad \text{(Note: } a \neq 0\text{)}$$

The term $b^2 - 4ac$ is called the discriminant; it can tell you how many solutions there will be:

> If $b^2 - 4ac > 0$, there are two real solutions.
> If $b^2 - 4ac = 0$, there is one real solution.
> If $b^2 - 4ac < 0$, there are two complex (imaginary) solutions.

Logarithms

$\log_k x = m$ such that $k^m = x$. Usually, $k = 10$.

Laws:

> $\log ab = \log a + \log b$
> $\log a/b = \log a - \log b$
> $\log a/b = -\log b/a$
> $\log a^n = n \log a$
> $\log \sqrt[n]{a} = 1/n \log a$

$\ln x$ (the natural log of x) $= m$ such that $e^m = x$.

Laws:

> $\ln e = 1$

Geometry

Triangle

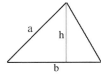

Perimeter of a triangle = $a + b + c$
Sum of the internal angles = $180° = \pi$ radians
Area of a triangle = $\dfrac{1}{2}\,bh$
Sides of a right triangle = $c^2 = a^2 + b^2$ (the Pythagorean theorem)

Parallelogram

Perimeter of a parallelogram = $2a + 2b$
Area of a parallelogram = $b \times h = ab\,\sin\alpha$
Volume of a box = $b \times h \times d$

Regular n-gon

This is an *n*-sided polygon with equal sides of length *b* and equal angles of $360°/n$ or $2\pi/n$ radians. See the following example:

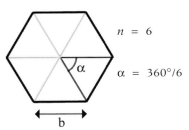

$n = 6$

$\alpha = 360°/6$

Perimeter = nb
Area = $\dfrac{1}{2}\,nb^2\,\cot(180°/n)$

Circle

Circumference of a circle = $2\pi r$
Area of a circle = πr^2
Surface area of a sphere = $4\pi r^2$
Volume of a sphere = $\dfrac{4}{3}\pi r^3$

Ellipse

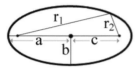 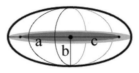

$r_1 + r_2$ = a constant
Area of an ellipse = πab
An ellipsoid is an ellipse rotated around the short axis (*b*).
Volume of an ellipsoid = $\dfrac{4}{3}\pi abc$

Right circular cylinder

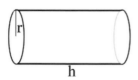

Surface area of a cylinder (not including the flat ends) = $2\pi rh$
Volume of a cylinder = $\pi r^2 h$

Half cone

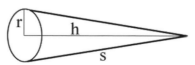

Surface area of a cone (not including the flat end) = πrs
Volume of a cone = $\dfrac{\pi r^2 h}{3}$

Torus

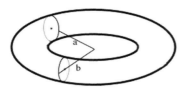

Surface area $= \pi^2(b^2 - a^2)$

Volume $= \dfrac{1}{4}\pi^2(a + b)(b - a)^2$

THE GEOMETRY OF EVERYDAY LIFE

TUNA SANDWICH   SNEAKER   GRANDMA

Trigonometry

For a right triangle:

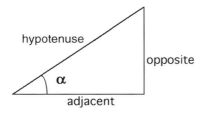

Function	Abbreviation	Definition	In Terms of sin and cos
Sine α	sin	$\dfrac{\text{opposite}}{\text{hypotenuse}}$	
Cosine α	cos	$\dfrac{\text{adjacent}}{\text{hypotenuse}}$	
Tangent α	tan	$\dfrac{\text{opposite}}{\text{adjacent}}$	$\dfrac{\sin}{\cos}$
Cotangent α	ctn	$\dfrac{\text{adjacent}}{\text{opposite}}$	$\dfrac{\cos}{\sin}$
Secant α	sec	$\dfrac{\text{hypotenuse}}{\text{adjacent}}$	$\dfrac{1}{\cos}$
Cosecant α	csc	$\dfrac{\text{hypotenuse}}{\text{opposite}}$	$\dfrac{1}{\sin}$

A mnemonic device for remembering the first three lines of this table is SOH CAH TOA: Sine is Opposite over Hypotenuse; Cosine is Adjacent over Hypotenuse; Tangent is Opposite over Adjacent. Some trigonometric equations follow.

One angle, α:

$$\sin^2 \alpha + \cos^2 \alpha = 1$$
$$\sin (-\alpha) = -(\sin \alpha)$$
$$\cos (-\alpha) = \cos \alpha$$
$$\sin (180° - \alpha) = \sin \alpha$$
$$\cos (180° - \alpha) = -\cos \alpha$$
$$\sin (90° - \alpha) = \cos \alpha$$
$$\cos (90° - \alpha) = \sin \alpha$$

Two angles, α and β:

$$\sin (\alpha + \beta) = \sin \alpha \cos \beta + \cos \alpha \sin \beta$$
$$\cos (\alpha + \beta) = \cos \alpha \cos \beta = \sin \alpha \sin \beta$$

THE ELECTROMAGNETIC SPECTRUM

Type of Radiation (Lowest to Highest Energy)	Frequency Range (Hz) $\nu = c/\lambda$	Wavelength Range (m) $\lambda = c/\nu$	
Radio	$<10^{11}$	$>10^{-3}$	↓ I N C R E A S I N G E N E R G Y
AM	$0.55 \times 10^6 - 1.60 \times 10^6$	$545 - 188$	
FM	$88 \times 10^6 - 108 \times 10^6$	$3.40 - 2.78$	
TV	$54 \times 10^6 - 890 \times 10^6$	$5.55 - 0.34$	
Radar	$10^9 - 10^{11}$	$3 \times 10^{-1} - 3 \times 10^{-3}$	
Infrared (heat)	$10^{11} - 3.8 \times 10^{14}$	$3 \times 10^{-3} - 8 \times 10^{-7}$	
Visible (light)	$3.8 \times 10^{14} - 7.5 \times 10^{14}$	$8 \times 10^{-7} - 4 \times 10^{-7}$	
Red	$3.95 \times 10^{14} - 4.76 \times 10^{14}$	$7.6 \times 10^{-7} - 6.3 \times 10^{-7}$	
Orange	$4.76 \times 10^{14} - 5.08 \times 10^{14}$	$6.3 \times 10^{-7} - 5.9 \times 10^{-7}$	
Yellow	$5.08 \times 10^{14} - 5.36 \times 10^{14}$	$5.9 \times 10^{-7} - 5.6 \times 10^{-7}$	
Green	$5.36 \times 10^{14} - 6.12 \times 10^{14}$	$5.6 \times 10^{-7} - 4.9 \times 10^{-7}$	
Blue	$6.12 \times 10^{14} - 6.67 \times 10^{14}$	$4.9 \times 10^{-7} - 4.5 \times 10^{-7}$	
Violet	$6.67 \times 10^{14} - 7.89 \times 10^{14}$	$4.5 \times 10^{-7} - 3.8 \times 10^{-7}$	
Ultraviolet	$7.5 \times 10^{14} - 3 \times 10^{17}$	$4 \times 10^{-7} - 10^{-9}$	
X-rays	$3 \times 10^{17} - 3 \times 10^{19}$	$10^{-9} - 10^{-11}$	↓
Gamma rays	$>3 \times 10^{19}$	$<10^{-11}$	↓

ν = frequency (s^{-1}); c = speed of light (2.99792458×10^8 m s^{-1}); λ = wavelength of light (m).

HALF-LIVES OF RADIONUCLIDES

Radionuclide	Half-life	Uncertainty	Units
^{3}H	4500	±8	days
^{14}C	5730		years
^{18}F	1.82951	±0.00034	hours
^{22}Na	950.97	±0.15	days
^{24}Na	14.9512	±0.0032	hours
^{32}P	14.262		days
^{33}P	25.34		days
^{35}S	87.51		days
^{44}Ti	22154	±456	days
^{45}Ca	162.61		days
^{46}Sc	83.831	±0.066	days
^{51}Cr	27.7010	±0.0012	days
^{54}Mn	312.028	±0.034	days
^{57}Co	272.11	±0.26	days
^{58}Co	70.77	±0.11	days
^{59}Fe	44.5074	±0.0072	days
^{60}Co	1925.12	±0.46	days
^{62}Cu	9.67	±0.03	min
^{65}Zn	244.164	±0.099	days
^{67}Ga	3.26154	±0.00054	days
^{75}Se	119.809	±0.066	days
^{85}Kr	3934.4	±1.4	days
^{85}Sr	64.8530	±0.0081	days
^{88}Y	106.626	±0.044	days
^{99}Mo	65.9239	±0.0058	hours
^{99m}Tc	6.00718	±0.00087	hours
^{109}Cd	463.26	±0.63	days
^{110m}Ag	249.950	±0.024	days
^{111}In	2.80477	±0.00053	days
^{113}Sn	115.079	±0.080	days
^{117m}Sn	14.00	±0.05	days
^{123}I	13.2235	±0.0019	hours
^{125}I	59.41	±0.13	days
^{125}Sb	1007.56	±0.10	days
^{127}Xe	36.3446	±0.0028	days
^{131}I	8.0197	±0.0022	days

HALF-LIVES OF RADIONUCLIDES (CONTINUED)

Radionuclide	Half-life	Uncertainty	Units
^{131m}Xe	11.934	±0.021	days
^{133}Ba	3853.6	±3.6	days
^{133}Xe	5.24747	±0.00045	days
^{134}Cs	753.88	±0.15	days
^{137}Cs	11015.	±20.	days
^{139}Ce	137.734	±0.091	days
^{140}Ba	12.7527	±0.0023	days
^{140}La	40.293	±0.012	hours
^{141}Ce	32.510	±0.024	days
^{144}Ce	284.558	±0.038	days
^{152}Eu	4945.5	±2.3	days
^{153}Gd	239.472	±0.069	days
^{153}Sm	46.2853	±0.0014	hours
^{154}Eu	3138.2	±6.1	days
^{155}Eu	1738.97	±0.49	days
^{166}Ho	26.7663	±0.0044	hours
^{169}Yb	32.0147	±0.0093	days
^{177}Lu	6.64	±0.01	days
^{181}W	121.095	±0.064	days
^{186}Re	89.248	±0.069	hours
^{188}Re	17.021	±0.025	hours
^{192}Ir	73.810	±0.019	days
^{195}Au	186.098	±0.047	days
^{198}Au	2.69517	±0.00021	days
^{201}Tl	3.0456	±0.0015	days
^{202}Tl	12.466	±0.081	days
^{203}Hg	46.619	±0.027	days
^{203}Pb	51.923	±0.037	hours
^{207}Bi	11523.	±19.	days
^{228}Th	698.60	±0.36	days

For radiation safety information, see:
http://www.orcbs.msu.edu/radiation/radsaf.html
http://www.uvm.edu/~radsafe/
http://www.umich.edu/~radinfo/introduction/index.htm
http://www.ehs.uiuc.edu/rss/ram/unitconvert.htm
Source: physics.nist.gov and http://www2.bnl.gov/ton/

FLUOROPHORES

Target	Fluorophore	Conditions of Measurement[a]	Ex Max (nm)	Em Max (nm)
	Alexa Fluor 350	Bound to protein*	347	442
	Alexa Fluor 430	Bound to protein*	434	540
	Alexa Fluor 488	Bound to protein*	495	519
	Alexa Fluor 532	Bound to protein*	531	554
	Alexa Fluor 546	Bound to protein*	556	573
	Alexa Fluor 568	Bound to protein*	579	604
	Alexa Fluor 594	Bound to protein*	591	618
	Alexa Fluor 633	Bound to protein*	632	647
	Alexa Fluor 647	Bound to protein*	650	668
	Alexa Fluor 660	Bound to protein*	663	690
	Alexa Fluor 680	Bound to protein*	679	702
	Allophycocyanin	Buf, pH 7.5	650	670
	AMCA		347	460
	Bodipy 493/503		500	506
	Bodipy 630/650	In methanol	630	650
	Bodipy 650/665	In methanol	650	665
	Bodipy TMR-X	Buf, pH 7.2	542	574
	Bodipy TR-X	Buf, pH 7.0	589	617
	Bodipy-FL	In methanol	505	513
	Calcein	Buf, pH 9.0	494	517
	Carboxyrhodamine 6G	Buf, pH 7.0	525	555
	Cascade Blue	Bound to protein*; Buf, pH 7.0	400	420
	Cascade Yellow	Buf, pH 8.0	402	545
	Chromomycin A3		458	526
	Cy3		550	602
	Cy5		650	680
	ELF 97 alcohol	Buf, pH 8.0	345	530
	Eosin	Buf, pH 8.0	524	544
	FITC	Bound to dextran; Buf, pH 8.0	494	518
	FITC	Buf, pH 9.0	494	518
	FITC	Bound to protein*; Buf, pH 8.0	494	518
	GFP		395	461
	JOE	Buf, pH 9.0	525	555
	LaserPro IR 790	Buf, pH 7.2	795	811
	Lissamine Rhodamine		570	590
	Lucifer Yellow CH	In water	428	536
	Magnesium Green	Buf, pH 7.05, 35 mM free Mg^{2+}	506	531
	Marina Blue	Bound to protein*; Buf, pH 8.0	365	460
	Merocyanine 540			
	NBD	In methanol	465	535
	Oregon Green 488	Bound to protein*; Buf, pH 8.0	496	524
	Oregon Green 514	Bound to protein*; Buf, pH 8.1	511	530
	Pacific Blue	Bound to protein*; Buf, pH 8.2	405	455
	perCP		470	529
	Resorufin	Buf, pH 9.0	570	585
	Rhod-2	Buf, pH 7.2; Ca^{2+} saturated	552	578
	Rhodamine 110	Buf, pH 7.0	496	520
	Rhodamine Green	Buf, pH 7.0	502	527
	Rhodamine Phalloidin	Buf, pH 7.0	555	580

FLUOROPHORES (CONTINUED)

Target	Fluorophore	Conditions of Measurement[a]	Ex Max (nm)	Em Max (nm)
	Rhodamine Red	Bound to protein*; Buf, pH 8.0	570	590
	Rhodol Green	Bound to protein*; Buf, pH 8.0	499	525
	R-phycoerythrin	Buf, pH 7.5	565	575
	Sodium Green	Buf, pH 7.0;135 mM Na$^+$	507	535
	Tetramethylrhodamine, 5-TAMRA	Buf, pH 7.0	550	575
	Tetramethylrhodamine, Rhodamine B	Bound to dextran; Buf, pH 7.0	550	575
	Tetramethylrhodamine, Rhodamine B	Bound to protein*; Buf, pH 8.0	550	575
	Texas Red	Bound to protein*; Buf, pH 7.0	595	615
	TRITC		554	611
	X-Rhodamine	Buf, pH 7.0	580	605
Calcium	Calcium Orange	Buf, pH 7.2; Ca^{2+} saturated	549	576
Calcium	Fluo-3	Buf, pH 7.2; Ca^{2+} saturated	506	526
Calcium	Fluo-4	Buf, pH 7.2; Ca^{2+} saturated	494	516
Calcium	Fura Red	Buf, pH 7.2; Ca^{2+} free	472	657
Calcium	Fura Red	Buf, pH 7.2; Ca^{2+} saturated	436	637
Calcium	Fura-2	Buf, pH 7.2; Ca^{2+} free	363	512
Calcium	Fura-2	Buf, pH 7.2; Ca^{2+} saturated	335	505
Calcium	INDO-1	Buf, pH 7.2; Ca^{2+} free	346	475
Calcium	INDO-1	Buf, pH=7.2; Ca^{2+} saturated	330	401
DNA	7-aminoactinomycin D	Bound to DNA	546	647
DNA	Acridine Orange	Bound to DNA	500	526
DNA	BOBO-1	Bound to DNA	462	481
DNA	BOBO-3	Bound to DNA	570	640
DNA	BO-PRO-1	Bound to DNA	462	481
DNA	DAPI	Bound to DNA	359	460
DNA	EthD-1	Bound to DNA	528	617
DNA	Ethidium bromide	Bound to DNA	518	605
DNA	Hoechst 33258	Bound to DNA	352	461
DNA	Hoechst 33342	Bound to DNA	352	461
DNA	POPO-1	Bound to DNA	434	456
DNA	PO-PRO-1	Bound to DNA	434	456
DNA	Propidium Iodide	Bound to DNA	536	617
DNA	SYTO Blue-40,-41, -42,-43	Bound to DNA	428±6	454±13
DNA	SYTO Blue-44, -45	Bound to DNA	450±5	478±7
DNA	SYTO Green-11,-14, -15,-20,-22,-25	Bound to DNA	515±7	543±13
DNA	SYTO Green-12,-13, -16,-21,-23,-24	Bound to DNA	494±6	515±7
DNA	SYTO Orange-80, -81,-82,-83	Bound to DNA	537±7	552±8
DNA	SYTO Orange-84,-85	Bound to DNA	567	583
DNA	SYTO Red-17,-59, -61,-64	Bound to DNA	615±15	632±13
DNA	SYTO Red-60,-62,-63	Bound to DNA	655±3	675±3
DNA	SYTOX Blue	Bound to DNA	445	470

FLUOROPHORES (CONTINUED)

Target	Fluorophore	Conditions of Measurement[a]	Ex Max (nm)	Em Max (nm)
DNA	SYTOX Green	Bound to DNA	504	523
DNA	SYTOX Orange	Bound to DNA	547	570
DNA	TOTO-1	Bound to DNA	514	533
DNA	TOTO-3	Bound to DNA	642	660
DNA	YOYO-1	Bound to DNA	612	631
DNA	YOYO-3	Bound to DNA	612	631
Lipid	Di-8-ANEPPS	Bound to phospholipid bilayer	468	635
Lipid	DiA	Bound to lipid	456	590
Lipid	DiD ($DiIC_{18}(5)$)	Bound to phospholipid bilayer	644	665
Lipid	DiI ($DiIC_{18}(3)$)	Bound to phospholipid bilayer	549	565
Lipid	DiO ($DiOC_{18}(3)$)	Bound to phospholipid bilayer	484	501
Lipid	DiR ($DiIC_{18}(7)$)	Bound to phospholipid bilayer	750	779
Lipid	ER Tracker Blue-White DPX	Bound to lipid	375	520
Lipid	FM 1-43	Bound to lipid	479	598
Lipid	FM 4-64	Bound to lipid	506	750
Lipid	Monobromobimane	Bound to lipid	394	490
Lipid	Nile Red	Bound to lipid	549	628
Lipid	RH 414	Bound to lipid	500	635
pH	BCECF	Buf, pH 5.5	482	520
pH	BCECF	Buf, pH 9.0	503	528
pH	Carboxy SNARF	Buf, pH 6.0	548	587
pH	Carboxy SNARF	Buf, pH 9.0	576	635
Potential	$DiBAC_4(3)$		493	516
Potential	$DiBAC_4(5)$		590	616
Potential	$DiIC_1$			
Potential	$DiOC_2$		484	500
Potential	$DiOC_5$		484	501
Potential	$DiOC_6$			
Potential	$DiSBAC_2(3)$		535	560
Potential	$DiSC_3$			
Potential	JC-1	Aggregate	593	595
Potential	JC-1	Monomer	498	525
Potential	JC-9	Aggregate	506	635
Potential	JC-9	Monomer	506	525
Potential	Oxonol V			
Potential	Oxonol VI			
Potential	Rhodamine 123	In methanol	507	529
Potential	Tetramethlyrhodamine, ethyl ester			
Potential	Tetramethylrhodamine, methy ester			
RNA	Acridine Orange	Bound to RNA	460	650
RNA	NeuroTrace 500/525 green Nissl	Bound to RNA	500	525

[a]Buf = In buffer; Ex Max = excitation maximum; Em Max = emission maximum.
*Protein is either IgG or BSA.
See http://www.probes/com and http://www.biorad.com.

FUNCTIONAL CHEMICAL GROUPS

Name	Formula	Comment
Acetyl	CH_3O	
Aldehyde	RCHO	Functional group is carbonyl; R is an H, alkyl, or aryl group
Alkane	C_nH_{2n+2}	Also known as aliphatic hydrocarbons
Alkene	C_nH_{2n}	Compounds with C=C functional groups
Alkyl	C_nH_{2n+1}	A group derived from an alkane minus one H
Alkyne	C_nH_{2n-2}	Compounds with C–C functional groups
Amino	$-NH_2$	
Aryl		Any group containing one or more fused benzene rings, less one H
Benzene	C_6H_6	Cyclic, with delocalized electrons
Bromo	-Br	Halogen
Carbonyl	-C=O	Functional group of aldehydes and ketones
Carboxyl	-COOH	Acids containing carboxyls are called carboxylic acids
Chloro	-Cl	Halogen
Ester	-COOR	R is an H, alkyl, or aryl group
Ethanol	CH_3-CH_2-OH	Produced by fermentation
Ethyl	$-CH_2$-CH_3	Alkyl
Fluoro	-F	Halogen
Hydroxyl	-OH	When part of a C-containing molecule, this defines an alcohol
Iodo	-I	Halogen
Ketone	RR'CO	Functional group is carbonyl; R and R' are alkyl and/or aryl groups
Methanol	CH_3OH	
Methyl	$-CH_3$	Alkyl
n-Butyl	$-CH_2$-CH_2-CH_2-CH_3	Alkyl
n-Propyl	$-CH_2$-CH_2-CH_3	Alkyl
Nitro	$-NO_2$	

GENETIC CODE

First position ↓	Second position →				Third position ↓
	T	**C**	**A**	**G**	
T	F Phe	S Ser	Y Tyr	C Cys	T
	F Phe	S Ser	Y Tyr	C Cys	C
	L Leu	S Ser	STOP	STOP	A
	L Leu	S Ser	STOP	W Trp	G
C	L Leu	P Pro	H His	R Arg	T
	L Leu	P Pro	H His	R Arg	C
	L Leu	P Pro	Q Gln	R Arg	A
	L Leu	P Pro	Q Gln	R Arg	G
A	I Ile	T Thr	N Asn	S Ser	T
	I Ile	T Thr	N Asn	S Ser	C
	I Ile	T Thr	K Lys	R Arg	A
	M Met	T Thr	K Lys	R Arg	G
G	V Val	A Ala	D Asp	G Gly	T
	V Val	A Ala	D Asp	G Gly	C
	V Val	A Ala	E Glu	G Gly	A
	V Val	A Ala	E Glu	G Gly	G

AMINO ACIDS

Amino Acid			Molecular Formula	MW
Alanine	Ala	A	CH_3-$CH(NH_2)$-COOH	89.09
Arginine	Arg	R	$HN=C(NH_2)$-NH-$(CH_2)_3$-$CH(NH_2)$-COOH	174.20
Asparagine	Asn	N	H_2N-CO-CH_2-$CH(NH_2)$-COOH	132.12
Aspartic acid	Asp	D	HOOC-CH_2-$CH(NH_2)$-COOH	133.10
Cysteine	Cys	C	HS-CH_2-$CH(NH_2)$-COOH	121.15
Glutamine	Gln	Q	H_2N-CO-$(CH_2)_2$-$CH(NH_2)$-COOH	146.15
Glutamic acid	Glu	E	HOOC-$(CH_2)_2$-$CH(NH_2)$-COOH	147.13
Glycine	Gly	G	NH_2-CH_2-COOH	75.07
Histidine	His	H	NH-CH=N-CH=C-CH_2-$CH(NH_2)$-COOH	155.16
Isoleucine	Ile	I	CH_3-CH_2-$CH(CH_3)$-$CH(NH_2)$-COOH	131.17
Leucine	Leu	L	$(CH_3)_2$-CH-CH_2-$CH(NH_2)$-COOH	131.17
Lysine	Lys	K	H_2N-$(CH_2)_4$-$CH(NH_2)$-COOH	146.19
Methionine	Met	M	CH_3-S-$(CH_2)_2$-$CH(NH_2)$-COOH	149.21
Phenylalanine	Phe	F	Ph-CH_2-$CH(NH_2)$-COOH	165.19
Proline	Pro	P	NH-$(CH_2)_3$-CH-COOH	115.13
Serine	Ser	S	HO-CH_2-$CH(NH_2)$-COOH	105.09
Threonine	Thr	T	CH_3-CH(OH)-$CH(NH_2)$-COOH	119.12
Tryptophan	Trp	W	Ph-NH-CH=C-CH_2-$CH(NH_2)$-COOH	204.23
Tyrosine	Tyr	Y	HO-p-Ph-CH_2-$CH(NH_2)$-COOH	181.19
Valine	Val	V	$(CH_3)_2$-CH-$CH(NH_2)$-COOH	117.15

See http://www.chemie.fu-berlin.de/chemistry/bio/amino-acids_en.html.

NEUTRAL RESIDUES OF AMINO ACIDS (AMINO ACIDS IN PROTEINS)

Amino Acid		Molecular Weight of Neutral Residue (g/mole)	Properties of Side Chain	Genetic Code
Alanine	A	71.1	Nonpolar	GCU, GCC, GCA, GCG
Cysteine	C	103.1	Nonpolar	UGU, UGC
Aspartic acid	D	115.1	Acidic	GAU, GAC
Glutamic acid	E	129.1	Acidic	GAA, GAG
Phenylalanine	F	147.2	Nonpolar	UUU, UUC
Glycine	G	57.0	Nonpolar	GGU, GGC, GGA, GGG
Histidine	H	137.1	Basic	CAU, CAC
Isoleucine	I	113.2	Nonpolar	AUU, AUC, AUA
Lysine	K	128.2	Basic	AAA, AAG
Leucine	L	113.2	Nonpolar	UUA, UUG, CUU, CUC, CUA, CUG
Methionine	M	131.2	Nonpolar	AUG
Asparagine	N	114.1	Polar, uncharged	AAU, AAC
Proline	P	97.1	Nonpolar	CCU, CCC, CCA, CCG
Glutamine	Q	128.1	Polar, uncharged	CAA, CAG
Arginine	R	156.2	Basic	CGU, CGC, CGA, CGG, AGA, AGG
Serine	S	87.1	Polar, uncharged	UCU, UCC, UCA, UCG, AGU, AGC
Threonine	T	101.1	Polar, uncharged	ACU, ACC, ACA, ACG
Valine	V	99.1	Nonpolar	GUU, GUC, GUA, GUG
Tryptophan	W	186.2	Nonpolar	UGG
Tyrosine	Y	163.2	Polar, uncharged	UAU, UAC

THE MITOCHONDRIAL GENETIC CODE

Codon	Nuclear	Mitochondria			
		Yeast	Drosophila	Plant	Mammal
UGA	STOP	Trp	Trp	STOP	Trp
AUA	Ile	Met	Met	Ile	Met
CUA	Leu	Thr	Leu	Leu	Leu
AGA	Arg	Arg	Ser	Arg	Arg
AGG	Arg	Arg	Ser	Arg	Arg

The genetic code of mitochondria varies with species and is different from the nuclear genetic code.

BUFFERS

Buffer	pK$_a$ at 25°C	pH range at 25°C
Oxalic acid (pK$_1$)	1.27	
H$_3$PO$_4$ (pK$_1$)	2.15	
Citric acid (pK$_1$)	3.13	
Oxalate$^-$ (pK$_2$)	4.27	
Acetic acid	4.76	
MES	6.1	5.5–6.7
NaHCO$_3$	6.35	
Bis-Tris	6.5	5.8–7.2
ADA	6.6	6.0–7.2
PIPES	6.76	6.1–7.3
ACES	6.8	6.1–7.5
MOPSO	6.9	6.2–7.6
Imidazole	7.0	6.2–7.8
BES	7.1	
MOPS	7.15	6.5–7.9
TES	7.4	6.8–8.2
HEPES	7.47	6.8–8.2
DIPSO	7.6	7.0–8.2
HEPPSO	7.8	7.1–8.5
POPSO	7.8	7.2–8.5
HEPPS (EPPS)	8.0	7.3–8.7
Tricine	8.1	7.4–8.8
Tris	8.1	7.0–9.0
Trizma	7.2–9.0	6.91–8.70
Bicine	8.26	7.6–9.0
Glycylglycine	8.4	7.5–8.9
TAPS	8.4	7.7–9.1
AMPSO	9.0	8.3–9.7
NH$_4^+$	9.25	
CHES	9.3	8.6–10.0
CAPSO	9.6	8.9–10.3
AMP	9.7	9.0–10.5
Glycine	9.78	
HCO$_3^-$ (pK$_2$)	10.33	
CAPS	10.4	9.7–11.1
Piperidine	11.12	
HPO$_4^{2-}$ (pK$_3$)	12.38	

PLUG AND CHUG

This section contains blank templates of frequently used calculations that are described in this book. Each is designed so that you can photocopy the page, then fill in the blanks (plug) and calculate (chug) the value you need, and then tape the result into your lab notebook.

Making a Solution of a Particular Molarity (See Chapter 4)

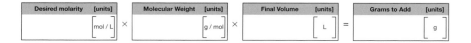

Making a Solution Using Hydrated Compounds (See Chapter 4)

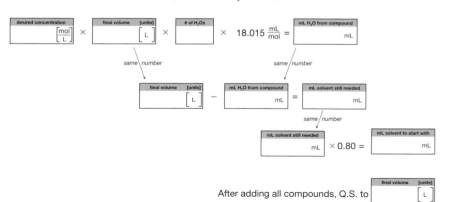

Diluting a Stock to a Particular Concentration (See Chapter 4)

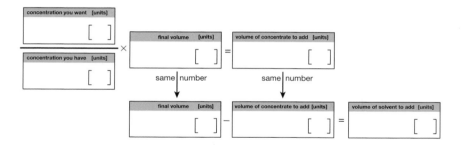

Nernst Equation (See Chapter 2)

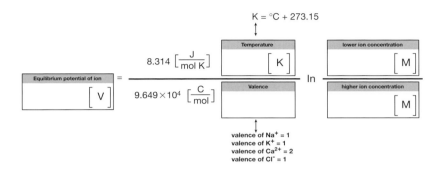

$$K = {}^{\circ}C + 273.15$$

valence of Na^+ = 1
valence of K^+ = 1
valence of Ca^{2+} = 2
valence of Cl^- = 1

Goldman Equation—To Calculate the Potential (V_m) Across a Biological Membrane (See Chapter 2)

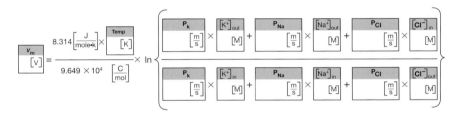

Permeability coefficients for ions in mammalian cells: $P_K = 5 \times 10^{-9} \frac{m}{s}$

$$P_{Na} = 5 \times 10^{-11} \frac{m}{s}$$

$$P_{Cl} = 1 \times 10^{-10} \frac{m}{s}$$

Counting Cells with a Hemocytometer (See Chapter 3)

Investigator_____

Date_____

Identity of cells to be counted:

_____

_____

_____

_____

Dilution(s) to make sample

1st: 1 + _____ for 1: [   ]

2nd: 1 + _____ for 1: [   ]

Notes:_____

_____

_____

_____

_____

_____

_____

_____

_____

_____

_____

_____

# of Cells   Chamber

[   ]

[   ]

[   ]

[   ]

[   ]

[   ]

[   ]

[   ]

[   ]

+ [   ]

[   ] Total Cell Count
(cells per µl sample)

×1000

[   ] Cells per ml sample

× [   ] 1st Dilution to make sample

× [   ] 2nd Dilution to make sample

[   ] Cell per ml

For an estimate of the uncertainty, find the mean ± the 95% confidence intervals of the ten counts, and multiply those two numbers by 1000 × 1st dilution factor × 2nd dilution factor. That will give you a range within which the actual number of cells per ml will be found.

Determining Percent Viability with a Hemocytometer
(See Chapter 3)

Investigator_____

Date_____

Identity of cells to be counted:

_____

_____

_____

Notes:_____

_____

_____

_____

_____

_____

_____

_____

_____

_____

_____

_____

_____

# Unstained		# Stained		Total Cells	Chamber
☐	+	☐	=	☐	
☐	+	☐	=	☐	
☐	+	☐	=	☐	
☐	+	☐	=	☐	
☐	+	☐	=	☐	
☐	+	☐	=	☐	
☐	+	☐	=	☐	
☐	+	☐	=	☐	
☐	+	☐	=	☐	
☐	+	☐	=	☐	

Unstained Cell Count	Stained Cell Count	Total Cell Count
☐	☐	☐

☐	÷	☐	×100 =	☐ %
Unstained Cell Count		Total Cell Count		% Viability

For a measure of uncertainty, calculate the % viability in each of the ten chambers, then calculate the mean and 95% confidence intervals of those ten numbers.

Preparation for Serial Dilution (See Chapter 4)

Final volume of each dilution: $V_f =$ [V_f [units]]

Concentration of most concentrated (initial) solution: $C_i =$ [C_i [units]]

Dilution factor: X = [X]

Number of dilutions: N = [N]

Required capacity of each container: [V_f [units]] + $\dfrac{[\,V_f\ \text{[units]}\,]}{[\,X\,] - 1}$ = [$V_{containers}$ [units]]

Label containers:

1st: $\dfrac{[\,C_i\ \text{[units]}\,]}{[\,X\,]}$ = [C_1 [units]]

2nd: $\dfrac{[\,C_i\ \text{[units]}\,]}{([\,X\,])^2}$ = [C_2 [units]]

3rd: $\dfrac{[\,C_i\ \text{[units]}\,]}{([\,X\,])^3}$ = [C_3 [units]]

Nth: $\dfrac{[\,C_i\ \text{[units]}\,]}{([\,X\,])^N}$ = [C_N [units]]

Amount of solvent to put into each container: $V_f =$ [V_f [units]]

Set pipet to: $\dfrac{[\,V_f\ \text{[units]}\,]}{[\,X\,] - 1}$ = [V_{pipet} [units]]

Determining the Concentration of a Solution of Pure Nucleotides Using Spectrophotometry (See Chapter 5)

Place to attach absorbance spectrum showing purity of solution.

$$\boxed{\text{Nucleotide Concentration}\left[\text{mM}\right]} = \frac{\boxed{A\lambda_{max}}}{\boxed{E_{NT}\left[\text{mM}^{-1}\text{cm}^{-1}\right]} \times \boxed{d\left[\text{cm}\right]}}$$

Nucleotide	pH	λ_{max} (nm)	E_{NT} (mM^{-1}cm^{-1})
dATP	7.0	259	15.4 mM^{-1}cm^{-1}
ATP	7.0	259	15.4 mM^{-1}cm^{-1}
dGTP	7.0	253	13.7 mM^{-1}cm^{-1}
GTP	7.0	252	13.7 mM^{-1}cm^{-1}
dCTP	2.0	280	13.1 mM^{-1}cm^{-1}
CTP	2.0	271	12.8 mM^{-1}cm^{-1}
dTTP	7.0	267	9.60 mM^{-1}cm^{-1}
UTP	7.0	262	10.0 mM^{-1}cm^{-1}

Determining the Concentration of a Solution of Oligonucleotides Using Spectrophotometry (See Chapter 5)

Place to attach absorbance spectrum showing purity of solution.

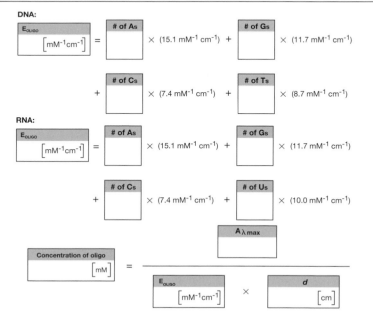

Determining the Concentration of an Uncontaminated Solution of Nucleic Acids Using Spectrophotometry (See Chapter 5)

Investigator:_____ Date: _____

λ_{max} = _____ nm

Place to attach absorbance spectrum showing purity of solution.

Concentration (c) of DNA or RNA $[\mu g/ml]$	=	A260	×	Conversion Factor $[\mu g/ml]$

Nucleic Acid	Conversion Factor
Single-stranded RNA	40 µg/ml
Single-stranded DNA	33 µg/ml
Double-stranded DNA	50 µg/ml

Determining Concentration Using Spectrophotometry

Investigator: _____ Date: _____

Measuring concentration of: _____

using the _____ spectrophotometer. λ_{max} = _____ nm

Concentration of Standard [Units]	Absorbance at λ_{max}

Standard curve for

Absorbance

2
1.5
1
0.5
0

Concentration of Standard

$$\text{Concentration of [units]} \quad [\quad] \quad = \quad \frac{\text{Absorbance} \quad - \quad \text{y – intercept}}{\text{Slope of Standard curve [units]} \quad [\quad]}$$

CONCENTRATIONS OF UNKNOWNS

Sample ID	Absorbance	Concentration of Sample [Units]

Index

CL

510
ADA